Notes on Numerical Fluid Mechanics and Multidisciplinary Design

Founding Editor

Ernst Heinrich Hirschel

Volume 155

Notes on Numerical Fluid Mechanics and Multidisciplinary Design publishes state-of-art methods (including high-performance methods) for numerical fluid mechanics, numerical simulation and multidisciplinary design optimization. The series includes proceedings of specialized conferences and workshops, as well as relevant project reports and monographs.

Indexed by SCOPUS, zbMATH, SCImago.

All books published in the series are submitted for consideration in Web of Science.

Ralf Heinrich
Editor

Advanced Aircraft Understanding via the Virtual Aircraft Model

Results of the German Research Initiative VitAM

Springer

Editor
Ralf Heinrich
Institute of Aerodynamics and Flow
Technology
German Aerospace Center (DLR)
Braunschweig, Germany

ISSN 1612-2909 ISSN 1860-0824 (electronic)
Notes on Numerical Fluid Mechanics and Multidisciplinary Design
ISBN 978-3-031-69424-0 ISBN 978-3-031-69425-7 (eBook)
https://doi.org/10.1007/978-3-031-69425-7

This Springer imprint is published by the registered company Springer Nature Switzerland AG
The registered company address is: Gewerbestrasse 11, 6330 Cham, Switzerland

Preface

The need for more economical, environmentally friendly and safer air traffic in the future represents a major challenge for the aviation industry. The national aviation industry and the closely intertwined European aviation industry are in direct competition with North American manufacturers in this respect. In fact, there is also a strong increase in competition from the BRICC countries (Brazil, Russia, India, China, Canada), with increased activity primarily in Canada and China. In order to maintain or even expand the position on the international market, better aircraft must be developed in shorter periods of time. This applies not only to new developments, but also to models derived from existing designs. The ability to describe an aircraft and its characteristics as accurately and comprehensively as possible in the computer, with associated simulation models in the sense of a virtual product, or more precisely in the form of a virtual model, plays a key role here. If the interaction of all relevant disciplines is represented in the model with suitable accuracy at an early stage, this also allows "virtual flight tests" in the computer long before the first flight of an aircraft type. This has the potential to identify wrong decisions in the design of an aircraft at an early stage in the design process and thus prevent high costs. A virtual aircraft model becomes equally important for achieving shorter development times and thus reduced costs as well. The physical tests currently required, in the wind tunnel as well as later in flight tests, will be significantly reduced by providing qualified models. This will also have an impact on the certification process. The project idea is based on the vision that an initial virtual aircraft model will be available at an early stage in the development process of an aircraft and will be increasingly refined in the course of development. In the long term, the virtual aircraft model should be an essential basis for future product definition and development and thus contribute to an improved product description and a shortened development process. The idea of the virtual aircraft model will be implemented in the project and demonstrated on industrially relevant use cases. The numerical models and simulation tools, whose development is the common core of all subprojects and thus forms the connecting intersection in the joint project, will be used for this purpose. The following partners are involved in the joint project with their use cases or contributions to use cases:

Airbus Operations:	**VitAir**—Virtual Airbus
Airbus Defence and Space:	**VitAMIn ABC**—Virtual Aircraft Model for the Industrial Assessment of Blended Wing Body Controllability
DLR:	**VitAM-Gust**—Virtual Aircraft Model for Gust Load Alleviation System Development
TU-Braunschweig ISM & WIRE:	**VitAM-Cal**—Virtual Aircraft Model Calibration—contributes to all use cases
TU-München LLB:	**VitAM-Flex**—Virtual Aircraft Model for Flexible Structures—contributes to VitAMIn ABC
Uni-Stuttgart IAG:	**VitAM–Turbulence**—contributes to VitAM-Gust

The reader will find separate chapters for each use case, for example "Use Case VitAM-Gust—Virtual Aircraft Model for Gust Load Alleviation System Development". All papers contributing to the use case are found in the respective chapter. The important results of the 4-year project initiative published in this book were also presented in the VitAM closing symposium which took place at the DLR Braunschweig, Germany, on 22nd of November, 2021.

Braunschweig, Germany
June 2023

Ralf Heinrich

Acknowledgements

The editor would like to express his thanks to all partners and colleagues who have contributed to the VitAM project in a very open and collaborative manner. The knowledge and engagement of each individual contributed to the success and high reputation of the project. The funding of partial activities through the Federal Ministry for Economic Affairs and Climate Action under contracts FKZ 20A1504A, 20A1504B, 20A1504C, 20A1504D, 20A1504E and 20A1504F in the framework of the air transport research program is gratefully acknowledged. Finally, thanks are due to B. Ernst for technical support in compiling this book. The editor is grateful to Prof. Dr. W. Schröder as the General Editor of the Notes on Numerical Fluid Mechanics and Multidisciplinary Design and to the Springer Verlag for the opportunity to publish the results of the research project.

Contents

Use Case VitAM-Gust and Subproject VitAM Turbulence

Overview of Sub-Project *VitAM-Gust*

Ralf Heinrich, Lars Reimer, Markus Ritter, and Britta Ernst

Abstract The development, testing and production of new aircraft are associated with considerable temporal and financial risks due to the product and manufacturing complexity. In order to accelerate the introduction of innovative technologies for more economical, more environmentally friendly and safer air vehicles and to better control the technological risks involved, DLR's guiding concept *The Virtual Product* aims at virtualizing the design, development and manufacturing processes, including the definition of an appropriate validation strategy. The availability of high-fidelity simulation tools with the capability to model multidisciplinary phenomena are a decisive part in that context. High demands are put on the numerical methods and solvers as the tools must provide reliable results in a robust manner for the entire flight envelope where nonlinearities are prominent in most of the disciplines. A key technology is the ability to enable *virtual flight testing* based on high-fidelity simulation tools. The progress made within the national research project *VitAM* (*Virtual Aircraft Model*) by the German Aerospace Center (DLR) in the recent past regarding the development, validation and demonstration of methods and tools to perform virtual flight tests is presented in this paper and the following ones of this book chapter. The present paper gives an overview of the DLR sub-project *VitAM-Gust*, the development of some important ingredients and simulation strategies required for virtual flight testing, as well as the simulation framework FlowSimulator used for their integration. Verification and validation aspects for the multi-disciplinary components as well as for the overall framework are addressed. As use case for the

R. Heinrich (✉) · L. Reimer · B. Ernst
DLR Institute of Aerodynamics and Flow Technology, Lilienthalplatz 7, 38108 Braunschweig, Germany
e-mail: ralf.heinrich@dlr.de

L. Reimer
e-mail: lars.reimer@dlr.de

B. Ernst
e-mail: britta.ernst@dlr.de

M. Ritter
DLR Institute of Aeroelasticity, Bunsenstraße 10, 37073 Göttingen, Germany
e-mail: markus.ritter@dlr.de

R. Heinrich (ed.), *Advanced Aircraft Understanding via the Virtual Aircraft Model*, Notes on Numerical Fluid Mechanics and Multidisciplinary Design 155,
https://doi.org/10.1007/978-3-031-69425-7_1

demonstration, virtual flight tests are performed for a medium size transport aircraft. The aerodynamic data achieved is then input for the development of a gust load alleviation system (GLA). The effectiveness of the GLA is proven via virtual flight tests involving aircraft gust encounters.

Keywords CFD · Multidiscipliary simulation · Virtual flight testing · Control surface modelling

1 Introduction and Goal of the Project

In order to meet the future social and economic challenges of aviation, the European body ACARE (Advisory Council for Aeronautics Research in Europe) has set extremely ambitious targets for future aircraft developments [1, 2]. For example, exhaust gases and perceived noise are to be reduced by more than 50 percent while simultaneously halving the ticket price. Furthermore, the accident rate is to be reduced by 80 percent and air traffic is to become largely independent of weather conditions. Currently, these goals are projected one step further into the year 2050 [3]. The aviation industry in Germany and Europe must find an answer to this and to the growing competition. Massive efforts of research and industry will be necessary to meet the requirements of significantly reduced environmental impact and simultaneously increasing transport capacity by means of new, forward-looking technologies. Numerical simulation based on higher-quality methods will play a key role here. The necessary skills and tools for numerically-based product development are to be developed and made available in a targeted manner in the *VitAM* joint project [4]. The DLR sub-project *VitAM-Gust* (*Virtual aircrft model for Gust load alleviation system development*) [5] is an essential element of the joint project and is oriented towards the objectives described in the overall project description. Three DLR institutes were involved as well as one small medium enterprise:

- DLR Institute of Aerodynamics and Flow Technology
- DLR Institute of Aeroelasticity
- DLR Institute of Flight Systems
- IBK Technology GmbH and Co. KG/ IBK Innovation GmbH and Co. KG.

To meet the challenges in *VitAM-Gust*, numerical simulation must make significant advances in the following areas:

- Improved flow simulation techniques:
  - improved modeling of detached flows (e.g. behind spoilers)
  - extension of the TAU code / Multidisciplinary Analysis (MDA) for use in system identification (SID) and parameter estimation

- Improved methods for structural simulation:
  - improved methods for automated preparation, adaptation and condensation of existing finite element (FE) structural models to modal dynamic master models (DMM)
- Flight mechanics methods:
  - development of flight control components for integration into the MDA simulation environment
  - adapt system identification and parameter estimation methods to use simulation data from MDA
- Integrated multidisciplinary simulation:
  - enable virtual flight tests including controller component
    - for pilot-induced maneuvers
    - for gust and wake vortex encounters
  - improved techniques for modeling moving control surfaces:
    - modeling of moving spoilers in high-fidelity CFD simulation on a realistic aircraft configuration (A320-ATRA)
  - improved robustness of free-flying aircraft simulation
- Improved reduced-order models (ROMs) and variable-fidelity methods for efficient aerodynamic database generation:
  - further development of industrial Aerodynamic Influence Coefficient (AIC) correction procedures for the Doublet Lattice Method (DLM)
- Development of processes for automated generation of aerodynamic databases based on virtual flight tests. This is an essential part of the *Virtual Aircraft Model* (VAM) under FlowSimulator:
  - further development, adaptation and implementation of necessary data structures and interfaces in the FlowSimulator data manager (FSDM)
  - enhancement of job control mechanisms and user interface
- Testing of the VAM on an industrial use case (see following chapter).

Future aircraft developments will only achieve necessary technological leaps with solutions for these problems. The *VitAM-Gust* project will make a decisive contribution to this as mentioned above. The idea of the VAM will be tested on an industrial use case and its potential demonstrated. This will be described in the following. But at first the project structure of *VitAM* is shown.

1.1 Project Structure

In Fig. 1 you can see the project structure of *VitAM*. In WP 1, the exact specification of what is to be captured, modeled and achieved with the VAM for the specific use case takes place. The targeted development of methods and processes takes place in WP 2, focused on the specific application. The setup of a VAM for the specific use case is worked on in WP 3. In WP 4, the verification and validation of the methods, the processes and the VAM as a whole are built and finally, in WP 5, the VAM is demonstrated on the specific use case. The WP's marked with red borders show where DLR and the subcontractor IBK were involved.

2 Use Case Definition of Sub-Project *VitAM-Gust* and Main Results

In the following, just the main results of *VitAM-Gust* will be described as well as the sub-project's use case. More detailed descriptions of the work accomplished can be found in subsequent papers of this book: [6] (methods for MDA modeling and application in the high-fidelity context), [7] (improvement of robustness of turbulence modeling), [8] (reduced order modeling with AIC methods), and [9] (methods for system identification for controller design). One topic which will also be described in greater detail in this paper, is the modeling of movable spoilers on a realistic transport aircraft (A320-ATRA), which will be of great interest in the near future.

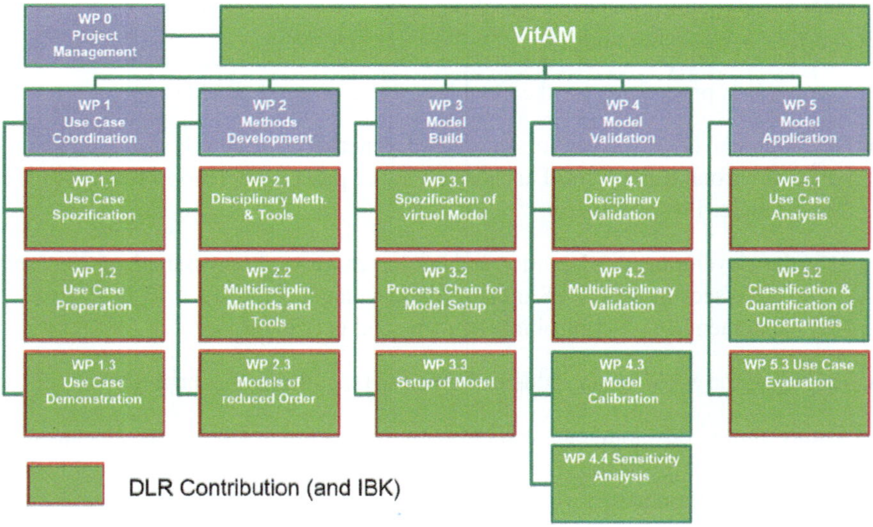

Fig. 1 Project structure of *VitAM*

2.1 *Use Case* VitAM-Gust

The application scenario is described below, see also Fig. 2. Essential part of the scenario was a concrete "development task", the development of a gust load alleviation system for load reduction and increase of passenger comfort. When an aircraft interacts with a vertical gust, it is usually assumed for certification purposes that the direction of flight is perpendicular to the gust front. This is also assumed in *VitAM-Gust*. In this case, three flight mechanical degrees of freedom are excited (movement in flight direction, vertical direction and pitch). The rudder does not need to be modeled. A reference simulation was performed first, for which the parameters below apply (CFD coupled to flight mechanics and structural dynamics—without controller).

The following flow parameters have been used:

- Incident flow data
  - Airspeed $Ma_\infty = 0.78$
  - Flight altitude $h = 37000\,ft$
  - Pressure $p_\infty = 21662.7\,N/m^2$
  - Temperature $T_\infty = 216.65\,K$
  - Density $\rho_\infty = 0.348395\,kg/m^3$
  - True airspeed $V_\infty = 230.133\,m/s$
- Gust parameters (FAR part 25.341)
  - Gust wavelength $\lambda = 213.36\,m$
  - Gust amplitude $U_{ds} = 10.52\,m/s$
- Aircraft A320-ATRA
  - Reference area $A_{ref} = 122.4\,m^2$
  - Reference mean chord $c_{ref} = 4.1935\,m$.

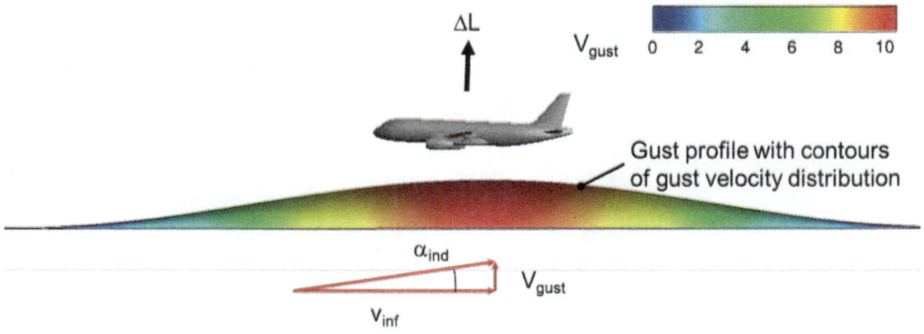

Fig. 2 *VitAM-Gust* use case: Interaction of A320 with a vertical gust

Once a suitable controller/load reduction system has been developed, the simulations should be repeated, then with a controller. The success of the controller used can then be evaluated in comparison to the uncontrolled reference state.

2.1.1 Components Required for Virtual Flight-Testing

For virtual flight testing, several simulation components are essential, see also Fig. 3. DLR's intention is to perform high-fidelity virtual flight tests. As CFD code for the aerodynamics, we make use of the DLR TAU Code [10], which is widely and successfully used at DLR, European universities and in aerospace industry. TAU makes use of the so called dual-mesh approach. The parallelization is based on domain decompositioning; MPI [11] is used as communication library. Many advanced turbulence models are implemented, such as Spalart-Allmaras, kω and even RSM-type models. Two central flux-discretization schemes can be used as well as several upwinding schemes. The code allows for arbitrary rigid-body and deforming single-mesh motions and, via the chimera technique (also known as overset method), also relative multi-mesh motions. This is of importance for the modeling of movable surfaces, especially for deployable control surfaces such as spoilers, see also paper [15] and below. To enable the simulation of atmospheric effects, the disturbance velocity approach (DVA) is used. It allows the simulation of gusts as well as of wake-vortices [14]. For CFD mesh generation purposes, the commercial grid generator CENTAUR [16] has been used. The test cases of the *VitAM-Gust* project involve several virtual flight tests of a medium range transport airplane. The aircraft is represented by coupled disciplinary simulation models for

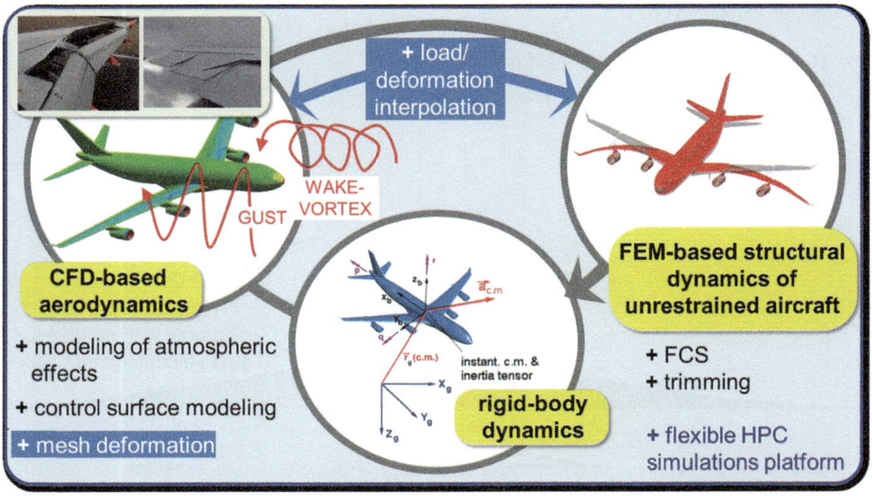

Fig. 3 *VitAM-Gust* simulation environment for virtual flight testing including atmospheric effects

aerodynamics, flight mechanics and structural dynamics. The CFD grid, consisting of roughly 8.5M nodes, representing the aerodynamics model is shown in Fig. 5. Control surface movements, e.g. of ailerons, elevators, are modelled in *VitAM-Gust* in a simplified fashion, disregarding the gaps occurring in reality between wing and deflected control surfaces. Mesh deformation is applied for deflecting the control surfaces, combined with a blending approach to bridge over the otherwise occurring gaps [17]. For trimming purposes, horizontal tail plane (HTP rotations are accomplished via mesh deformation, too, involving a no-normal-movement constraint imposed on the fuselage to prevent fuselage shape violations [17]. More information on control surface movement can be found below. Another important component in the process chain is the spatial coupling, meaning the transfer of aerodynamic loads to the structure and, vice versa, the projection of deformations from the structural mesh to the aerodynamic mesh. Details of the available algorithms to do so, like moving least squares-based, can be found in [17]. For propagating CFD surface mesh deformations into the CFD volume mesh, on which the CFD solver operates, a mesh deformation method based on RBF (radial basis functions) is available in TAU [18] and FlowSimulator [19], as well as a module based on linear or nonlinear elasticity [24]. The methods could also be combined, to improve the robustness and mesh quality of the deformed mesh, see also [17]. The structural modeling is based on FEM-based structural dynamics of free-flying aircraft. Usually the model is condensed to a modal model, which simplifies and accelerates the simulation. Before you can start a virtual flight test, the aircraft needs to be trimmed. The control surfaces and the thrust have to be adapted, such that the sum of all forces and moments are zero. For doing so, a dedicated module has been developed. A complex trim example with six degrees of freedom and engine simulation can be found in [17] (one engine out scenario). For modeling of rigid body dynamics, including elastic deformations, the module FSDynafly has been developed, to compute the aircraft's reaction to aerodynamic forces (including engine thrust), gravity and inertia loads. Based on a high-fidelity virtual flight test a simple flight control system (FCS) has been developed with the aim to reduce the loads during a vertical gust encounter and to improve passenger comfort.

Simulation Environment

An important part of the simulation system is the software integration environment. The FlowSimulator suite is chosen as central HPC-ready software integration framework. Its development, aiming for harmonization of CFD workflows and for making it a mainstay of CFD-based multidisciplinary simulations, has been initiated more than a decade ago by Airbus and several European research facilities (including DLR) and has been ongoing ever since.

FlowSimulator is organized in three layers, see Fig. 4: The *control layer* is the end user's main access layer; it is where the process scripts reside. Python scripting is used for rapid process development and integration of diverse software components written in different programming languages. The FSDataManager (FSDM)—the core of FlowSimulator—is the *data layer*. In line with HPC requirements, it

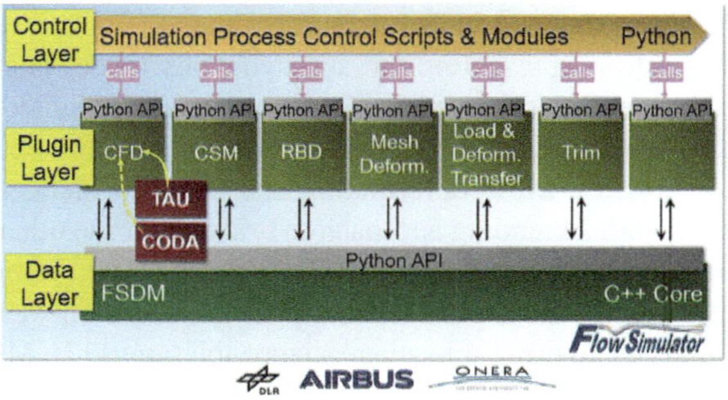

Fig. 4 Software architecture of the FlowSimulator suite

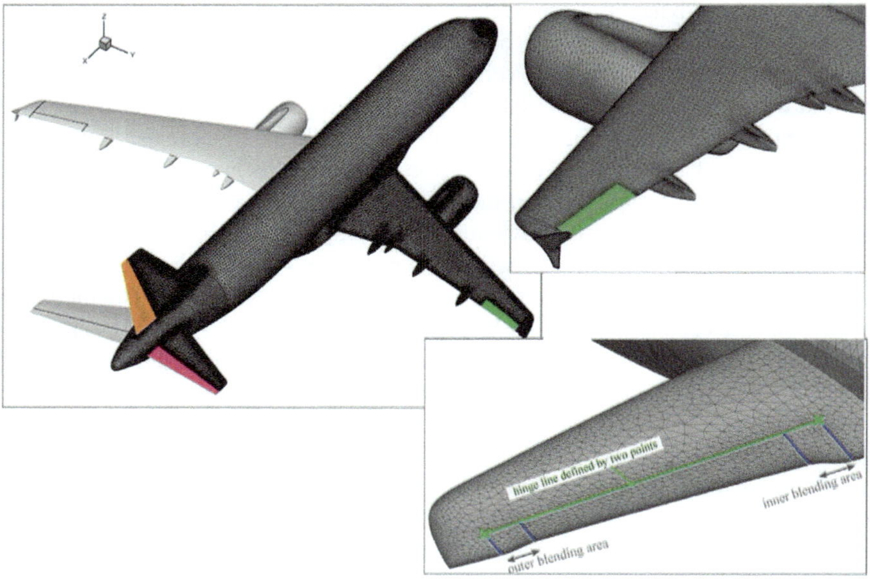

Fig. 5 Overview of the CFD surface mesh used in *VitAM-Gust* simulation for virtual flight testing including atmospheric effects. The movable control surfaces are marked in red, orange and green. In *VitAM-Gust* just the elevator and HTP have been used for the gust encounter simulations. The lower right image illustrates the blending concept applied when deflecting the control surfaces relevant in *VitAM-Gust* solely via mesh deformation means

enables in-memory data exchange and is designed fully parallel from the ground up, supporting existing intra-field and (optionally) inter-field parallelization based on MPI. The FSDM provides data containers for unstructured and structured multi-block meshes, CAD geometry, and hierarchical descriptions of configuration and simulation scenario, as well as tools for mesh selection and mesh-to-mesh interpolation. Implementations for parallel multi-dimensional arrays are available, classes for vectors, matrices, coordinate systems, affine transformations, binary search operations, etc. Interfaces to graph-based mesh partitioning libraries exist, as well as to common linear algebra packages. For performance reasons, the FSDM is written in C++; the majority of classes is wrapped to Python. The FSDM library is provided under the LGPL v2 open source license; hence, others may become involved in the development and benefit from the joint effort. The *plugin layer* comprises the set of all applications/tools interfacing with the FSDM and being available for use in FlowSimulator, e.g. CFD and CSM solvers, modules for volume mesh deformation, spatial coupling, flight mechanics, trim and other purposes.

The philosophy of FlowSimulator is that the plugins must not communicate data with other plugins but only with the FSDM. That way, the risk of process chains becoming dependent on specifics of individual process components should be avoided, making process chains easier to maintain and extend, e.g., replacing present CFD and/or CSM solvers by more advanced candidates. For more information on FlowSimulator and its use for multidisciplinary simulation purposes, consult [20–23].

Deploying and Retracting Spoilers

Building on preliminary work from projects $C^2A^2S^2E$, *ComFliTe* and *Digital-X*, a fair work share of the *VitAM* and the DLR *VicToria* project was spent on integrating dynamically deploying and retracting spoilers in the CFD simulation and eventually in virtual flight tests. Spoilers are of special interest from a modeling perspective due to their importance for the flight control system, for load alleviation measures for instance. The aerodynamic characteristics of spoilers, however, are challenging to predict, even with RANS-based simulation tools. The reason is: spoiler deflections always induce flow separation. The outer flow field, and so the response of the aircraft, however remains largely unaffected until the spoiler wake extends beyond the trailing edge of the wing. This interaction is responsible for the well-known nonlinear response and time delays associated to spoiler controls. In addition, when talking about simulating movable spoilers, it needs to be ensured that the simulation tool is capable of capturing the transition from fully retracted state, i.e. the spoiler being flush with the rest of the wing surface, to deflected state with deflection angles of up to 50 degree or beyond. Due to the large range in deflection angles, the simulation of moving spoilers typically utilizes an overset grid technique. It turned out that the initially chosen steady simulation approach, using the negative version of the one-equation turbulence model by Spalart and Allmaras (here labelled as SA-neg [13]), already is well capable of reproducing the experimental pressure distribution for static spoiler deflection up to 50 degree. However, quantitative differences remained for all investigated turbulence models, with each of them predicting a too low base pressure aft of the deployed spoiler. Nonetheless, this deficiency could be overcome if the simulation is performed time-accurately instead. Details on the validation work done by DLR concerning moving spoilers can be found in [30, 31].

In addition, based on successful work on rather simple generic non-deforming 2D and 3D configurations and based on above findings, attempts have been made in

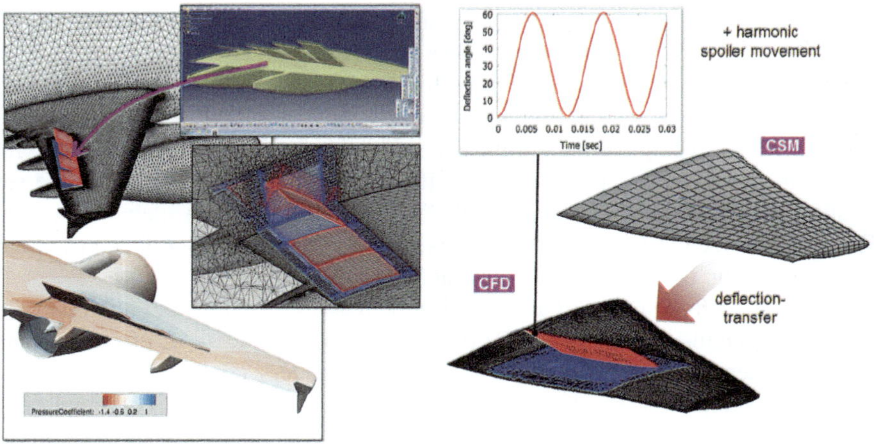

Fig. 6 Modeling of movable spoilers at A320 (*left*) and prototype tests regarding the deformation transfer in CFD/CSM-coupled analyses of wings with moving spoilers (*right*)

VitAM and *VicToria* to also integrate movable spoilers into more realistic configurations (here: A320). The left Fig. 6 illustrates the steps performed: (1) CAD modeling of spoilers and associated spoiler cavity in the wing; (2) generation of component meshes (shown here: a 3-block approach, consisting of an unstructured background mesh including the aircraft, a structured cavity mesh, and structured spoiler meshes); (3) assembly of the overall Chimera mesh (involving dry runs with moving spoilers). The final image shows a sample steady simulation result (c_p distribution) with the roll spoilers being deflected by 60, 50, 40 and 30 degrees (inboards to outboards). One important question in this context was: How small needs the gap to be so that the load distribution for the configuration with integrated but retracted spoilers is almost identical to that of the clean configuration, i.e. the configuration without installed spoilers? Early simulations with 10 mm gaps (aerodynamic mean chord of A320 is 4.1935 m) showed significant differences in the pressure distribution, especially upstream of the spoilers (Fig. 7, top). In the end, it turns out that gaps down to 1 mm are necessary to achieve the targeted result, both in terms of local pressure distribution and integral coefficients (Fig. 7, bottom).

Investigations on different meshing concepts have shown the following: Although the control of the mesh overlap with the approaches based on structured component meshes is much easier with respect to the mesh setup, the convergence behavior (in TAU) with purely unstructured meshes is more favorable. In the former case, especially the turbulence equation's residual stagnates (SA-neg is used), whereas in the latter case a converged solution is obtained (Fig. 8).

By the end of the project, it was possible to successfully simulate the complete unsteady deployment and retraction process of the roll spoilers on the A320 using a gap width of 1mm. Figure 9 shows snapshots of some instants in time during the spoiler deployment. The spoilers were deflected from 0 to 40 degrees and fully

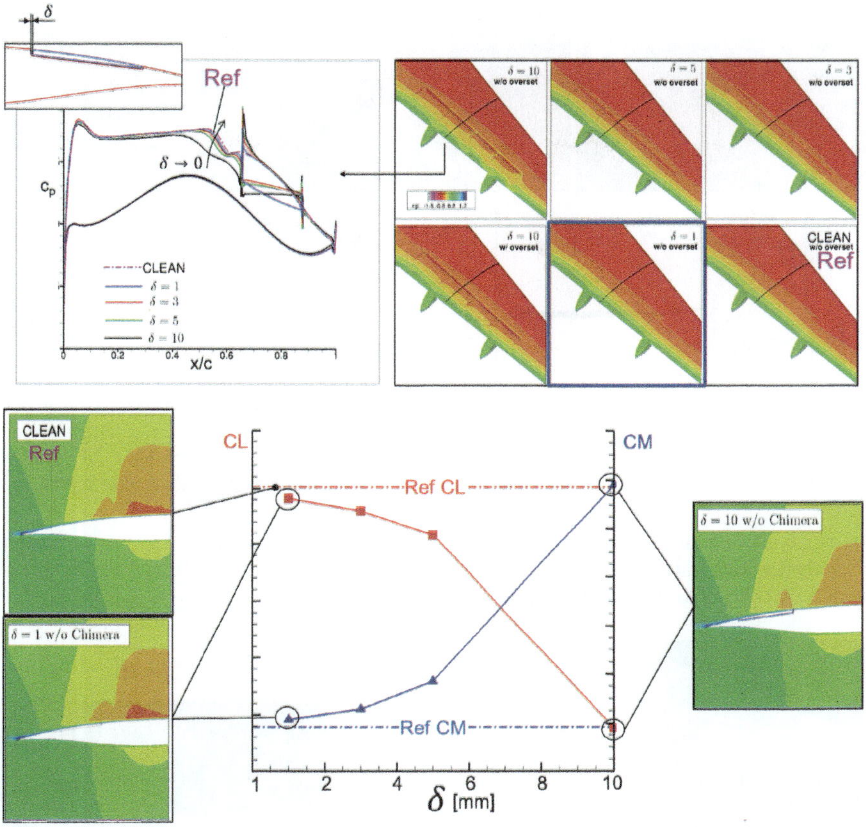

Fig. 7 Effect of the gap size between installed (but fully retracted) spoilers and the rest of the wing surface on surface pressure distribution (*top*) and integral aerodynamic coefficients (*bottom*) (TAU simulations of A320 at $Ma_\infty = 0.78$, h = 37000 ft, $\alpha = 0°$, SA-neg turbulence model)

retracted again in two seconds. In terms of the overall mesh resolution, a rather coarse CFD mesh was used. It is tuned for unsteady flight test simulations; also, some geometry details are simplified or removed (e.g. strakes). The half-model mesh has 72 million points. Many of them are placed in the spoiler area just to meet the requirements for the spoiler integration via overset grids. Currently, in preparation for virtual flight tests for elastic aircraft, moving spoilers are tested in interaction with deforming structures (Fig. 6, right).

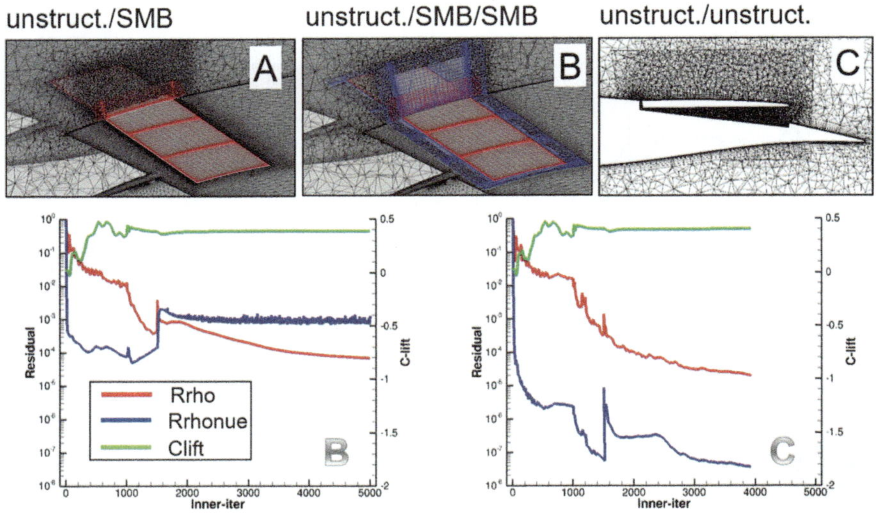

Fig. 8 Studies on the best meshing strategy for spoiler integration and the effects on CFD solver convergence (simplified A320 configuration; $Ma_\infty = 0.78$, h = 37000 ft, $\alpha = 0°$; SA-neg turbulence model; SMB refers to structured multi-block meshes)

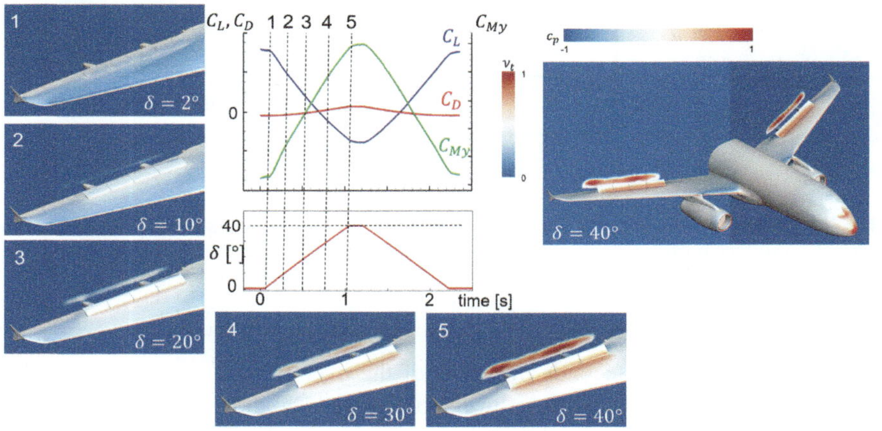

Fig. 9 Unsteady simulation of deploying and retracting roll spoilers at the A320 ($Ma_\infty = 0.78$, h = 37000 ft, $\alpha = 0°$, SA-neg turbulence model)

2.2 *Verification, Validation and Application*

Accurate initial conditions are indispensable for maneuver simulations to ensure a steady straight non-accelerated level flight if no control surface inputs were applied. Otherwise, the force and moment residuals from a not perfectly converged trim simulation would lead to small drifts during the unsteady simulation which would

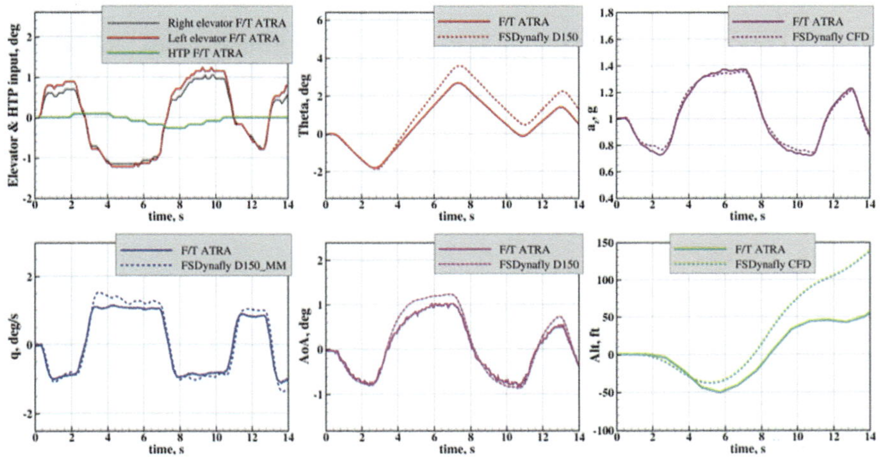

Fig. 10 Comparison of flight test data recorded during a "13211-longitudinal-maneuver" to simulation results of corresponding virtual flight tests

then superimpose the maneuver response and lead to errors. So, the level flight was the first verification test case. The HTP and thrust have to be adjusted such that the sum of all forces and moments are zero. This is done by the trim module. To test if the trim process was workingproperly, a follow-on unsteady simulation was started. It could be shown, that the aircraft flew exactly leveled and unaccelerated, so that the trim process was proven to be correct. A comparison with flight test data was carried out as a validation test. With the A320 ATRA a so called "13211-longitudinal-maneuver" was performed. Starting from level flight the HTP and the elevator were moved in a certain way. This maneuver was carried out in identical manner in the virtual flight test with the simulation system developed. The results are shown in Fig. 10. The solid lines belong to the flight test, the dashed lines to the virtual flight test. The upper left image shows the input signals for the deflections of the control surfaces as measured during the ATRA flight tests, green for the HTP, grey for the right elevator and red for the left elevator; all values are stated in degrees. Exactly the same input was used for the virtual flight test. The image below plots the pitching rate q in degrees per second. The match between flight test and virtual flight test is good especially considering the uncertainties existing in the recorded flight test data. The same holds for the pitching angle theta and angle of attack (AoA), both plotted in degrees, see the images in the second row of Fig. 10. The acceleration in z-direction, normalized by gravity acceleration g, matches the flight test data very well, see upper right figure. In the figure below, the change of altitude in feet is displayed. In this case, the match is not as good, which is not understood since the acceleration in z-direction is predicted very accurately. All in all, the *VitAM-Gust* team was very satisfied with the results.

During the course of the project, further validations and verifications as well as numerous virtual flight tests for system identification (SID) were carried out. The

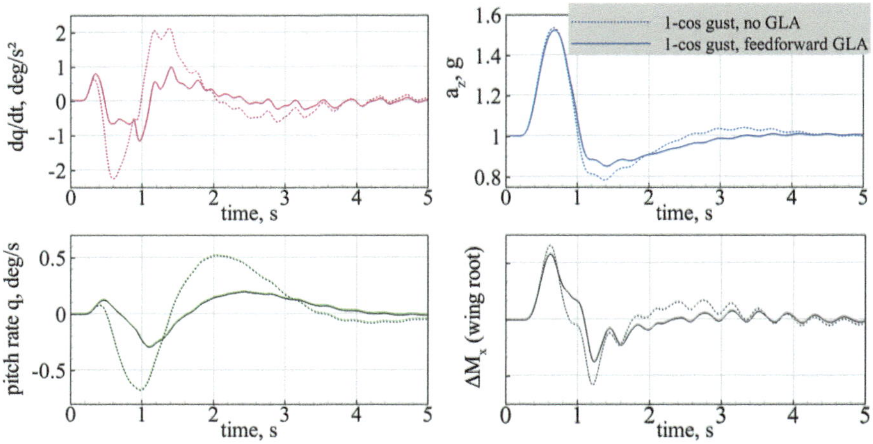

Fig. 11 Comparison of virtual flight test data of a gust encounter scenario with and without GLA system

resulting data is relevant for the design of the gust load alleviation (GLA) system. Further information can be found in the publications of this book, especially in [6], or in [25].

Next, a selected highlight result is presented in brief, which stems from virtual flight tests with gust encounters, simulated once with and once without the GLA system. The associated flow conditions and gust parameters are stated in Sect. 2.1. The results are presented in Fig. 11. The dashed lines represent the results *without* GLA system; the solid lines are the ones *with* GLA system. On the left, the pitch rate and the pitch acceleration are shown. With GLA, both values are significantly reduced. This is beneficial for the passenger's comfort. On the right, the acceleration in vertical direction and the root wing bending moment are presented. The latter is slightly reduced in case of an active GLA. Unfortunately, nearly no effect is visible for the acceleration in z direction. Generally speaking, the GLA system works though its effectiveness still needs to be improved. Nevertheless, it is shown that a GLA system can be successfully designed solely on the basis of virtual flight test data, which is considered a great success.

3 Summary

In the *VitAM-Gust* project, an MDA environment for performing virtual flight tests was further developed and improved. The essential disciplines are considered: Aerodynamics, structural dynamics, and flight mechanics including trim. An industrially relevant test case was used to demonstrate the functionality and potential of the MDA environment. In the test case, a GLA system for the DLR research aircraft

A320 ATRA has been developed on the basis of high-fidelity virtual flight tests. Furthermore, the MDA environment was validated against measurement data from real flight tests. From the pool of flight test data, a "13211 longitudinal maneuver" was selected, where the HTP and the elevator are deflected according to a certain pattern. The deflections of the control surfaces were measured during flight test and used as input for the corresponding simulations. The match between real and virtual flight test are good.

For the development of the flight controller, virtual maneuvers were calculated using specially designed inputs for the control surfaces that could not be applied in real flight tests. This allowed the necessary simulation times to be significantly reduced [9]. Based on the obtained aerodynamic data (the VAM), a GLA system was developed using system identification methods. To test it, virtual flight tests were performed with and without GLA system. A gust encounter of a long-wave gust was calculated, as it is also used in the certification process of a real aircraft. By means of the GLA system, the pitch rate and pitch acceleration could be significantly reduced, which is beneficial to passenger comfort. All in all, it was shown that GLA systems or flight controllers can be successfully developed solely on the basis of high-fidelity virtual flight tests. Therefore, the *VitAM-Gust* project is deemed a success. Furthermore, the transient deployment and retraction of roll spoilers on a realistic full-aircraft model, the A320 ATRA, was demonstrated in an unsteady CFD simulation, which is also considered a great success.

Acknowledgements As part of the *VitAM* project the partial funding of this research under contract FKZ 20A1504A of the Federal Ministry for Economic Affairs and Climate Action in the framework of the air transport research program is gratefully acknowledged.

References

1. European Aeronautics, A Vision for 2020, Report of the Group of Personalities (2021)
2. Luftfahrt 2020, Die deutsche Luftfahrtforschung, Partner im globalen Wettbewerb, Bericht der Arbeitsgruppe Luftfahrtforschung mit einer Empfehlung des Beirats Luftfahrtforschung, 30. Mai (2001)
3. Flightpath 2050: Europe's Vision for Aviation, Maintaining Global Leadership and Serving Society?s Needs, Report of the High Level Group on Aviation Research, ISBN 978-92-79-19724-6, European Union (2011)
4. Vorhabensbeschreibung zum Verbundvorhaben, Advanced Aircraft Understanding via the Virtual Aircraft Model - VitAM (2015)
5. Vorhabensbeschreibung zum Eigenantrag des DLR Virtual Aircraft Model for Gust Load Alleviation System Development - VitAM (2015)
6. Ritter, M., Reimer, L.: Multidisciplinary, High-Fidelity Simulation for Identification of Steady and Dynamic Transport Aircraft Derivatives, to be published 2024
7. Togiti, V., Braun, S.: Robustness improvements for differential Reynolds stress modeling using a logarithmic formulation, to be published 2024
8. Kaiser, C., Friedewald, D.: Aeroelastic AIC-based Reduced Order Model with CFD-Corrections for Gust Encounter Simulations, to be published 2024

9. Roeser, M., Mönnich, W.: System Identification from Virtual Flight Test Data for a Flexible Transport Aircraft, to be published 2024
10. D. Schwamborn, T. Gerhold, R.H.: The DLR TAU-Code: Recent Applications in Research and Industry. In: Proceedings of "European Conference on Computational Fluid Dynamics" ECCOMAS CDF 2006, Delft The Netherland (2006)
11. http://www.mpi-forum.org/docs/docs.html
12. Spalart, P.R., Allmaras, S.R.: A one-equation turbulence model for aerodynamic flows. AIAA Paper **92–0439**, 1992 (1992)
13. Allmaras, S.R., Johnson, F.T., Spalart, P.R.: Modifications and clarifications for the implementation of the Spalart-Allmaras turbulence model. In: Proceedings of 7th International Conference on Computational Fluid Dynamics (ICCFD7), 9–13 July, 2012, Waimea, HI, USA (2012)
14. Heinrich, R., Reimer, L.: Comparison of different approaches for modeling atmospheric effects like gusts and wake-vortices in the CFD code TAU. In: 17th INTERNATIONAL FORUM ON AEROELASTICITY AND STRUCTURAL DYNAMICS (Forum Proceedings), pages 1-13.IFASD 2017 - 17th international forum on aeroelasticity and structural dynamics, 25.-28. Juni 2017, Como, Italien (2017). ISBN 978 88 97576 28 0
15. Streher, L.B., Heinrich, R.: Modellierung beweglicher Steuerflächen militärischer Konfigurationen: ein Vergleich zwischen numerischen Methoden und CFD-Lösern, DLRK Dresden, 29.09.2022 - Diabolo-Sondersitzung (2022)
16. https://www.centaursoft.com/
17. Reimer, L., Heinrich, R., Geisbauer, S., Leicht, T., Görtz, S.: Virtual Aircraft Technology Integration Platform: Ingredientsfor Multidisciplinary Simulation and Virtual Flight Testing, 2021 AIAA SciTech Forum. Accessed from 11–15 and 19–21 Jan
18. Heinrich, R., Kroll, N., Neumann, J., Nagel, B.: Fluid-structure coupling for aerodynamic analysis and design - a DLR perspective. In: 46th AIAA Aerospace Sciences Meeting and Exhibit 2008, Reno, paper 2008-056 (2008)
19. Barnewitz, H., Stickan, B.: Improved Mesh Deformation. In: Eisfeld, B. et al. (eds.) Management and Minimisation of Uncertainties and Errors in Numerical Aerodynamics. Notes on Numerical Fluid Mechanics and Multidisciplinary Design, vol 122, pp. 219–243. Springer (2013)
20. Meinel, M., Einarsson, G.O.: The FlowSimulator framework for massively parallel CFD applications. PARA Conf., 6–9 June, 2010, Reykjavik, Iceland (2010)
21. Reimer, L. et al.: Multidisciplinary Analysis Workflow with the FlowSimulator. In Benoit, C. (ed.) Proceedings of the Onera Scientific Day 2012—CFD Workflow: Mesh, Solving, Visualizing. 3 Oct, 2012, Paris, France (2012)
22. Reimer, L. et al.: The FlowSimulator-A software framework for CFD-related multidisciplinary simulations. NAFEMS Conf.: Computational Fluid Dynamics (CFD) - Beyond the Solve. 2-3 Dec, 2015, Munich, Germany (2015)
23. Reimer, L.: Multidisciplinary analysis workflow with the flow simulator. In: Benoit, C. (ed.) Proceedings of the Onera Scientific Day 2012-CFD Workflow: Mesh, Solving, Visualizing. 3 Oct, 2012, Paris, France (2012)
24. Rempke, A.: Netzdeformation mit Elastizitätstheorie in multidisziplinärer FlowSimulator-Umgebung. DGLR Symposium of STAB, 9-11 Nov, 2016, Braunschweig, Germany (2016)
25. Ritter, M.R., Handojo, V., Krüger, W.R., Roeser, M.S., Mönnich, W., Kirmse, T., Reimer, L., Heinrich, R., Geisbauer, S., Görtz, (2021) Virtual Aircraft Technology Integration Platform: From Virtual Flight Testing towards Simulation-based Certification. In: AIAA Scitech 2021 Forum. AIAA SciTech 2021 Forum, Virtual Conference (2021). Accessed from 11–21 Jan 2021. https://doi.org/10.2514/6.2021-1203
26. Meinel, M., Einarsson, G.O.: The flow simulator framework for massively parallel CFD applications. In: PARA Conference, Reykjavik, Iceland (2010). Accessed from 6–9 June 2010
27. Reimer, L. et al.: Multidisciplinary analysis workflow with the flow simulator. In: Benoit, C. (ed.) Proceedings of the Onera Scientific Day 2012-CFD Workflow: Mesh, Solving, Visualizing. Paris, France (2012). Accessed from 3 Oct 2021

28. Reimer, L. et al.: The flowsimulator-A software framework for CFD-related multidisciplinary simulations. NAFEMS Conf.: Computational Fluid Dynamics (CFD) - Beyond the Solve. Munich, Germany (2015). Accessed from 2–3 Dec 2015
29. Reimer, L., Heinrich, R., Ritter, M.R.: Towards higher-precision maneuver and gust loads computations of aircraft: status of related features in the CFD-based multidisciplinary simulation environment flowsimulator. In: 21st STAB/DGLR Symposium on New Results in Numerical and Experimental Fluid Mechanics. 142, Springer. 21st STAB/DGLR Symposium, Darmstadt, Germany (2019). Accessed from 6–7 Nov 2018
30. Geisbauer, S., Löser, T.: Towards the Investigation of Unsteady Spoiler Aerodynamics. AIAA-2017-4229, 35th AIAA Applied Aerodynamics Conference, Denver, p. 2017. CO, USA (2017). Accessed from 5–9 June 2017
31. Geisbauer, S.: Numerical Simulation and Validation of the Aerodynamics of Static and Dynamic Spoilers. AIAA-2020-2776, AIAA Aviation Forum (2020). Accessed from 15–19 June 2020

High-Fidelity Simulation for Identification of Steady and Dynamic Transport Aircraft Derivatives

Markus Ritter and Lars Reimer

Abstract This work describes an approach to identify flight mechanic characteristics of a flexible jet transport aircraft based on high fidelity aerodynamics. A parameter identification process uses time histories of a multidisciplinary flight dynamics simulation including rigid body and elastic degrees of freedom together with the time dependent control surface inputs to estimate steady and dynamic derivatives. The focus is on the identification of the longitudinal parameters of the aircraft for which the elevator and the horizontal tail plane are deflected with wavelet-based inputs of selected frequencies and amplitudes. The total flight time is in the order of a few seconds only, appropriate modification of the inertia properties of the aircraft further reduces the computational costs. The approach for the multidisciplinary flight dynamics simulations are described in detail, including the methods for the high fidelity, CFD-based aerodynamics, the flight mechanics, and the structural dynamics. The test case is a narrow-body jet airliner for which the simulation models and the results of two flight dynamics simulations for system identification as well as a gust load alleviation scenario are presented.

Keywords CFD · Numerical aeroelasticity · Numerical flight dynamics · Virtual flight testing · Simulation by certification · DLR-ATRA

M. Ritter (✉)
German Aerospace Center (DLR) Institute of Aeroelasticity, Göttingen, Germany
e-mail: markus.ritter@dlr.de

L. Reimer
German Aerospace Center (DLR) Institute of Aerodynamics and Flow Technology, Braunschweig, Germany
e-mail: lars.reimer@dlr.de

R. Heinrich (ed.), *Advanced Aircraft Understanding via the Virtual Aircraft Model*, Notes on Numerical Fluid Mechanics and Multidisciplinary Design 155,
https://doi.org/10.1007/978-3-031-69425-7_2

1 Motivation

The identification of the flight mechanic stability and control characteristics is a decisive part of the aircraft design and certification process. As soon as the aircraft enters experimental testing, a series of time and cost consuming test flights, almost exclusively with multistep, one-surface-at-time inputs for each flight condition, is conducted. This of course requires that the aircraft physically exists, which means that testing takes place late. Substantial time and cost savings are feasible by the application of state of the art multidisciplinary simulation methods comprising the relevant flight physical disciplines, at least aerodynamics, flight mechanics, and structural dynamics. The multidisciplinary simulation can be applied as soon as the necessary simulation models are available in the design process. In this work we focus on high fidelity analysis, which means that aerodynamic loads are calculated by CFD and structural deformations by a finite element method. Thus a CAD model of the aircraft is required for the CFD grid generation process. The structural (finite element) model is available from the structural optimization (sizing) process much earlier in the design. In general, the single disciplines can be represented by independent simulation tools with arbitrary fidelity (which can be selected according to the desired accuracy). Theoretically, an multidisciplinary simulation program can be implemented in a completely integral manner by using a common discretization of the partial differential equations (e.g. finite elements or finite volumes) in space and time. However, this approach is rarely used in practice because very customized simulation models and tools of the single disciplines are used to setup the models (structural models of complex transport aircraft are exclusively represented by finite element models, compressible and viscous flows at transonic airspeeds are typically simulated by finite volume methods). Thus it is common practice for multidisciplinary analysis to couple independent programs. The coupling methods must transfer data between the programs, for instance aerodynamic loads must be mapped conservatively onto the structural model (which has a different topology) and the structural displacements must be interpolated onto the aerodynamic grid. The particular methods used for these tasks are detailed in the subsequent sections.

In the context of this work, we refer to the multidisciplinary, high fidelity flight dynamics simulation as *Virtual Flight Test*. As an extension to previous publications of the authors in this field [12], we propose a method to further reduce the simulation time and computational costs by altering particular parameters of the aircraft model, here we modify the aircraft's inertia properties (moment of inertia about the lateral axis, I_{yy}) to increase the maneuverability.

The real flight test to identify the flight mechanic stability and control characteristics of the aircraft typically uses multistep, one-surface-at-time inputs for different flight conditions to excite the aircraft in its expected flight mechanic eigenmodes. Inputs such as doublets, *1-3-2-1-1* and sinusoidal signals (mostly with increasing amplitude) are comparatively easy to execute manually by experienced pilots and are therefore used as the basis for classical system identification maneuvers. Such input signals are exemplarily plotted in Fig. 1. An a-priori aerodynamic and flight mechanic

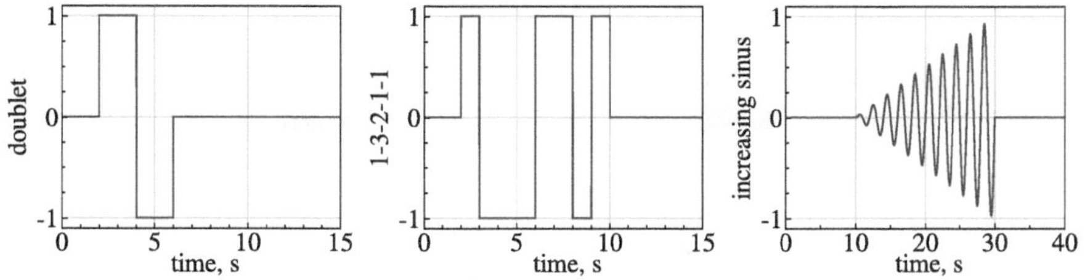

Fig. 1 Typical input signals for manual (pilot) actuation of primary control surfaces for aircraft identification: Doublet, *1-3-2-1-1*, and sinus with increasing amplitude. Execution times are within the range of seconds

model of the aircraft is used for the design of the input signals; the models are typically derived from wind tunnel testing or low- and mid-fidelity aerodynamic analysis. Because of uncertainties in the a-priori aircraft models, frequencies in a certain range above and below the expected eigenfrequencies of the flight mechanic modes must be considered for the maneuver design.

Frequency sweeps are an alternative type of input for system identification. It is possible to obtain a complete frequency response of the system to a single control surface input from a single flight test point. However, each control surface must be excited separately and the length of the excitation might become comparatively long (up to 2 min) which renders this approach costly.

Maneuvers in which multiple control surfaces are excited simultaneously offer the potential for significant reduction of the flight test time and costs. A new kind of multi-axis input signals based on the wavelet transform has been designed and applied to system identification by Roeser [1]. The signals are parametrized by their frequency contents and the times at which the particular frequencies are excited with as few parameters as possible. The wavelet transform yields a *Time-Frequency Representation* (TFR) of a signal; the signal can be uniquely reconstructed from its TFR. However, the development of such signals is an involved task. The frequency content must be chosen carefully, especially if frequency proximity of rigid-body and elastic degrees of freedom (e.g. short period mode and low frequency wing bending) is present, which is often the case for large jet transport aircraft. Of course such complex multi-axis inputs can only be used in numerical simulations and not in real flight tests if the excitation is done by a pilot. The goal of the project *VitAM-Gust* in terms of aircraft identification is focused on the longitudinal characteristics, thus only inputs which excite the longitudinal motion are applied. For the system identification process developed for the DLR project *VicToria*, multi-axis inputs with excitation of both the longitudinal and the lateral degrees of freedom of the aircraft were used [12]. For this work, the input signals excite the following primary control surfaces: Ailerons (symmetric deflections only), elevators, and horizontal tail plane. So far, spoilers cannot be deflected as function of time during the maneuver due to the complexities involved in the chimera technique, but this will be addressed in future applications. The thrust setting is kept constant since only small variations of

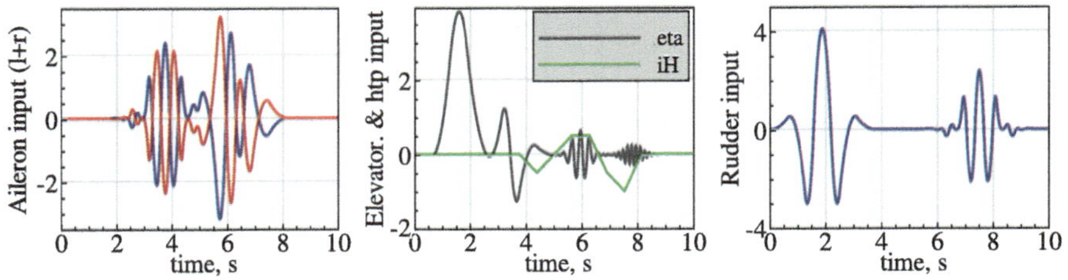

Fig. 2 Wavelet-type input signals for multi-axis actuation of primary control surfaces for virtual flight tests [12]

the flight speed compared to the trim speed are expected during the maneuver. Due to the high computational costs of the time-stepping CFD procedure, a particular goal for the development of the multi-axis wavelet inputs was to reduce the simulation time as much as possible. For the virtual flight tests of the longitudinal maneuvers of this work, a signal with a length of 3.6 s and one with only 1.6 s was developed. As an example, the multi-axis signals developed for the project *VicToria* are plotted in Fig. 2 (the longitudinal control surface inputs for the virtual flight tests of this work are presented in Sect. 4.1). The subsequent system identification process, in which the particular flight mechanic characteristics of the aircraft are identified from the results of the virtual flight tests, is detailed in Chap. 5 of this book (Roeser and Mönnich, *System Identification from Virtual Flight Test Data for a Flexible Transport Aircraft*).

2 Numerical Methods for the Virtual Flight Test

In this section, the methods for the virtual flight test are described. The virtual flight test corresponds to a multidisciplinary numerical simulation for which aerodynamics, flight mechanics, and structural dynamics are coupled in space and time. The aircraft is in free-flight and the flexibility of the airframe is taken into account. A steady trim simulation is performed before the actual unsteady maneuver simulation in order to obtain equilibrium of loads to ensure a steady straight level flight state if no control surface inputs were applied. The unsteady maneuver simulation is then performed in discrete time domain where the deflection of the control surfaces are prescribed and the coupled governing equations of motion (including rigid-body and elastic degrees of freedom) are solved for each time step. All quantities of interest necessary for the subsequent system identification process are derived from the flight dynamics and structural dynamics states in postprocessing.

2.1 Flight Dynamics Equations of Motion

The flight dynamics vector equations of motion (EOMs) for the free-flying flexible aircraft used for this work are detailed in literature [2–4]. Although only longitudinal dynamics are considered for *VitAM-Gust*, the equations of motion with all degrees of freedom (longitudinal and lateral) are presented here. Even in the numerical simulation (virtual flight test), the full set of equations is solved, but with control surface inputs restricted to the excitation of longitudinal motions. The well-known *mean axes* conditions, which enable a simple decoupling of rigid-body and elastic DOFs [3], are used together with a modal approach for the calculation of the structural deflections in generalized coordinates. The motion of the aircraft in space is described with respect to a geodetic, earth-fixed frame, and a body-fixed frame, both depicted in Fig. 3. A set of Euler angles, $\underline{\boldsymbol{\Phi}}$, is used to define the attitude of the body frame within the geodetic frame [24], angular rates of the body frame resolved in the body frame are denoted by vector $\boldsymbol{\Omega}_b$, and the velocity components along the particular axes of the body-fixed frame are denoted as $\boldsymbol{V}_b$,

$$\underline{\boldsymbol{\Phi}} = [\, \Phi \;\, \Theta \;\, \Psi \,]^T \tag{1a}$$

$$\boldsymbol{\Omega}_b = [\, p \;\, q \;\, r \,]^T \tag{1b}$$

$$\boldsymbol{V}_b = [\, u \;\, v \;\, w \,]^T \tag{1c}$$

The structural model of the aircraft is depicted in Fig. 3, where the discrete mass elements of the lumped mass model are denoted by dm_i. The nodal displacement $\boldsymbol{u}_i$ and its velocity, $\dot{\boldsymbol{u}}_i$, of a mass point dm_i due to structural deformation are calculated by the modal approach using a superposition of the eigenvectors of the structure, ${}^{p}\boldsymbol{\Phi}^{i}$:

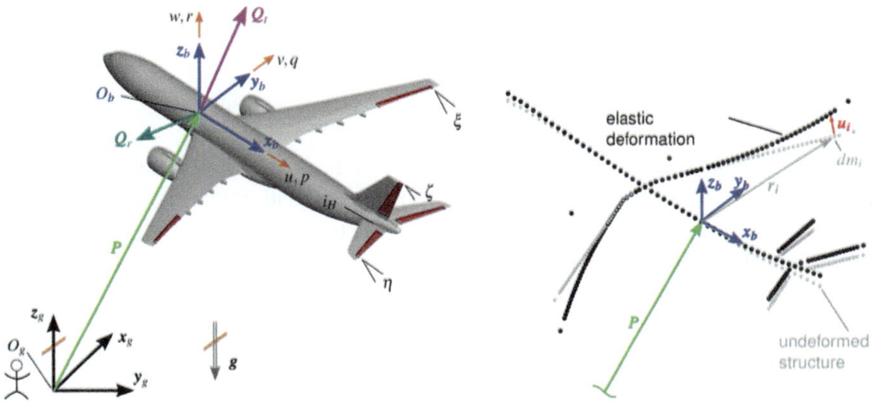

Fig. 3 Inertial and body-fixed coordinate frames (left). Lumped mass model of the airframe with discrete mass point dm_i in undeformed position, r_i, and displacement due to structural deformation, u_i (right)

$$\boldsymbol{u}_i = {}^p\boldsymbol{\Phi}^i \, q_p \tag{2a}$$

$$\dot{\boldsymbol{u}}_i = {}^p\boldsymbol{\Phi}^i \, \dot{q}_p \ , \quad (p = 1, \ldots, m) \tag{2b}$$

With the index m denoting the number of eigenvectors. The vector equations of motion used for the flight dynamics simulations in this work comprise the translational and rotational rigid-body motion as well as the elastic structural deformations. They can be derived from mechanical principles such as Lagrange's equations of the second kind [2–5]. Only the final EOMs are given in the following.

The linear momentum equation of motion is given as

$$\mathring{\boldsymbol{V}}_b m + \mathring{\boldsymbol{\Omega}}_b \times \overline{\boldsymbol{rm}} + \boldsymbol{\Omega}_b \times \left[\boldsymbol{V}_b m + \boldsymbol{\Omega}_b \times \overline{\boldsymbol{rm}} \right] - \left(\boldsymbol{M}_{bg} \boldsymbol{g} \right) m = \boldsymbol{Q}_t \ . \tag{3}$$

A ring above a variable denotes a temporal derivative with respect to the body-fixed frame of reference. The term $\overline{\boldsymbol{rm}}$ considers the inertial coupling between rotational and translational momentum, it becomes zero if the origin of the body-fixed frame is located at the center of mass of the undeformed structure. The entire mass of the aircraft is denoted by m and $\boldsymbol{g}$ denotes the gravity vector, which is resolved in the geodetic frame. Matrix $\boldsymbol{M}_{bg}$ transforms the gravity vector from the geodetic into the body-fixed frame.

The rotational momentum equation of motion is given as:

$$\boldsymbol{J}\mathring{\boldsymbol{\Omega}}_b - \mathring{\boldsymbol{V}}_b \times \overline{\boldsymbol{rm}} + \boldsymbol{V}_b \times \left[\boldsymbol{\Omega}_b \times \overline{\boldsymbol{rm}} \right] + \boldsymbol{\Omega}_b \times \left[\boldsymbol{J}\boldsymbol{\Omega}_b - \boldsymbol{V}_b \times \overline{\boldsymbol{rm}} \right] = \boldsymbol{Q}_r \ , \tag{4}$$

where $\boldsymbol{J}$ denotes the tensor of inertia of the undeformed structure. In general, the tensor of inertia can be expressed as function of the structural deformations, but it is kept constant in this work for simplicity.

Assuming a linear structural model, the governing equation of the elastic deformation for mode p is given as

$$M_{pk} \, \ddot{q}_k + \Omega^k \, q_k = Q^p \ , \tag{5}$$

where M_{pk} denotes entry (p, k) of the generalized mass matrix (for mass normalized eigenvectors, the generalized mass matrix reduces to the identity matrix), and Ω^k the eigenvalue (generalized stiffness) of the k-th structural eigenvector. To simplify the numerical integration, the hybrid (first and second order in time) EOMs presented above are formulated as a system of coupled, nonlinear ordinary differential equations (ODEs) [5].

The nonconservative generalized forces and moments for translational and rotational momentum as well as for the structural work are denoted as $\boldsymbol{Q}_t$, $\boldsymbol{Q}_r$, and Q^p, respectively. In this work, the generalized forces of the translational and angular momentum are composed of aerodynamic and propulsive (thrust) forces and moments only:

$$\boldsymbol{Q}_t = \boldsymbol{R}^A + \boldsymbol{F} \ , \tag{6a}$$

$$\boldsymbol{Q}_r = \boldsymbol{Q}^A + \boldsymbol{Q}^F \ . \tag{6b}$$

Here, $\boldsymbol{R}^A$ denotes the resulting aerodynamic forces acting on the aircraft with respect to the axes of the body-fixed frame, and $\boldsymbol{F}$ denotes the resulting propulsive forces, also resolved in the body-fixed frame. Accordingly, $\boldsymbol{Q}^A$ are the resulting aerodynamic moments, and $\boldsymbol{Q}^F$ denotes the propulsive moments, both resolved in the body-fixed frame. The generalized forces of the structural governing equation are calculated by:

$$Q^p = {}^p\boldsymbol{\Phi}_0{}^T \boldsymbol{f}^s \ , \tag{7}$$

where $\boldsymbol{f}^s$ denotes the aerodynamic forces transferred to the structural nodes. The aerodynamic forces are calculated by the CFD solver from the pressure distribution and provided at the nodes of the CFD grid (vertex-centered integration); to obtain equivalent forces on the structural nodes, a linear mapping based on an interpolation matrix is used for the (globally conservative) force transfer:

$$\boldsymbol{f}^s = \boldsymbol{H}^T \boldsymbol{f}^a \ . \tag{8}$$

The interpolation matrix H is based on the evaluation of either two- or three-dimensional *radial basis functions* (RBF), where a variety of suitable RBFs can be applied, for example ones with local or global support [6, 7].

2.2 Aerodynamic Method

As mentioned in the introduction, a crucial part of the multidisciplinary analysis is the method used for the calculation of the aerodynamic forces. To obtain highly accurate results, as few assumptions and simplifications as possible are made regarding the fluid dynamics equations to be solved and the modeling of the geometry of the aircraft. Thus methods based on the solution of the potential equations (such as VLM or DLM) are not considered but a CFD method based on the solution of the RANS and URANS equations is applied. For this work, the CFD solver TAU of the DLR, which is a collection of modules for the solution of the compressible Euler and Navier-Stokes equations, is employed. Complex geometries can be modeled by hybrid, unstructured grids and several turbulence models are available together with multigrid algorithms to accelerate convergence. An ALE formulation enables time-domain solutions with moving meshes for concurrent rigid-body and elastic motions. The code is developed by DLR and validated extensively [8]. To simplify multidisciplinary computations on clusters, a python-based framework enables the modular connection of external programs, such as structural and flight dynamics solvers or mesh deformation methods by efficient parallel data structures [9]. For the virtual flight tests presented in this work, the TAU code is used in the RANS (for steady trim) and URANS (for unsteady maneuver) mode with the Spalart-Allmaras turbulence model.

The DLR TAU code offers an interface, the *Motion Module*, to prescribe translational and rotational rigid body motions of the body-fixed frame of reference (which is likewise joint to the CFD grid). For the translational rigid-body motion, the three components of the position and velocity vector resolved along the axes of the geodetic frame of reference must be specified. For the rotational motion, the attitude (Euler angles) $\underline{\boldsymbol{\Phi}}$ and the body rates $\boldsymbol{\Omega}_b$ as defined in Eq. (1) are specified. Both translational and rotational rigid-body motions are converted to equivalent velocities of the CFD grid nodes by the Motion Module. Additionally, elastic structural deformations and velocities, $\boldsymbol{u}_i$ and $\dot{\boldsymbol{u}}_i$, as defined in Eq. (2), must be considered. The structural velocities are calculated from past grid node locations via a second-order accurate backward difference scheme.

2.3 Trim Simulations for Steady Straight Horizontal Flight

Accurate initial conditions are indispensable for the maneuver simulations to ensure a steady straight level flight state if no control surface inputs were applied. Otherwise, the force and moment residuals from a not perfectly converged trim simulation would lead to small drifts during the unsteady simulation which would then superimpose the maneuver response and lead to errors in the subsequent system identification. For the solution of the trim problem (steady straight level flight), the pitch angle, the angle of the htp, and the thrust must be calculated to obtain force and moment equilibrium of the aircraft. In this work, the equations for the calculation of the force and moment balance are formulated with respect to the body-fixed frame of reference. A modified Newton-Rhapson method is used for root-finding with a Jacobian matrix evaluated by a first-order finite difference approach. The Jacobian matrix is calculated at the beginning of the trim simulation and kept constant since structural deformations and aerodynamic nonlinearities are expected to be small for the aircraft configuration investigated. The structural deformations are not trimmed explicitly, but an iterative static coupling sub-loop ensures a converged aeroelastic solution during each iteration of the Newton method. Based on the flight dynamics EOMs described above, the trim problem to be solved for is defined in terms of constraints on particular variables as

$$\begin{bmatrix} \mathring{\boldsymbol{V}}_b \\ \mathring{\boldsymbol{\Omega}}_b \\ \ddot{q}_p \end{bmatrix} \stackrel{!}{=} \begin{bmatrix} \boldsymbol{0} \\ \boldsymbol{0} \\ 0 \end{bmatrix}; \quad \boldsymbol{\Omega}_b \stackrel{!}{=} \boldsymbol{0}; \quad \dot{q}_p \stackrel{!}{=} 0; \quad \dot{\boldsymbol{P}}_g = \begin{bmatrix} V_{TAS} \\ 0 \\ 0 \end{bmatrix}, \tag{9}$$

where V_A denotes the (trim) air speed in flight direction, which coincides with the x direction of the geodetic system. The last expression (the constraint on $\dot{\boldsymbol{P}}_g$, which is the velocity of the aircraft in the geodetic frame) forces the aircraft to a steady horizontal, straight flight with respect to the geodetic system.

2.4 *Inputs and Outputs for the System Identification Process*

The system identification requires the inputs (primary control surface inputs–elevator, ailerons, horizontal tail plane–as functions of time) and the outputs (flight dynamic and elastic states, also as functions of time) of the virtual flight test [1]. The inputs are known in advance, the flight dynamic and elastic states are the results of the solution of the equations of motion described in Sect. 2.1. Table 1 lists selected outputs and inputs from the virtual flight tests which used for the subsequent system identification process described in Chap. 5 of this book (Roeser and Mönnich, *System Identification from Virtual Flight Test Data for a Flexible Transport Aircraft*). Although only longitudinal maneuvers are the object of this paper, all states, i.e. including longitudinal and lateral, are given here. The states and results are converted from the coordinate systems shown in Fig. 3 to the usual flight mechanic coordinate system according to *ln9300* standard (body frame x axis positive in flight direction, z axis positive downwards) in order to avoid confusion and to simplify the subsequent system identification process.

3 Simulation Models: CFD Aerodynamic, FEM Structural, and Couplingmodel

The test cases of the *VitAM-Gust* project are three virtual flight tests of a mid-range jet transport aircraft. An aerodynamic and a structural simulation model represent the aircraft. The aerodynamic model consists of the CFD grid which consists of approximately 8.5 million nodes and 25 million cells. The outer shape and the CFD surface grid are based on surfaces derived from a high quality CAD model of the aircraft. A particular problem with CFD simulations of jet transport aircraft is the modeling of engines and nacelles. For simplicity, often through-flow nacelles are used (as has been done in previous works of the authors [10, 18]). Using engine boundary conditions yields significant improvements in the CFD simulation; therefore, the static pressure at the inlet (Fan) disk and the differences in the total pressure and total temperature are prescribed at the outlet disk, respectively. Thrust is generated by the total pressure difference and available as an "aerodynamic force" (static pressure distribution integrated over the in- and outlet boundaries of the engines). This method was used for the simulations presented in the following section.

The virtual flight test is based on particular control surface inputs. Accurate deflections of control surfaces in CFD simulations is a complex task and object of current research. Sophisticated approaches based on Chimera and sliding or patched grid

Table 1 Selected outputs (and inputs) from the virtual flight tests used for the system identification process [1]

Output	Unit	Description
Time	s	Simulation time
$\dot{\boldsymbol{V}}_K = [\dot{u}_K, \dot{v}_K, \dot{w}_K]^T$	m/s^2	Time derivative of flight-path velocity
$\boldsymbol{V}_K = [u_K, v_K, w_K]^T$	m/s	Flight-path velocity
$xPos, yPos, h$	m	Position of aircraft in geodetic frame, altitude
$\boldsymbol{V} = [u, v, w]^T$	m/s	Velocity in body frame
$\dot{\boldsymbol{\Omega}}_b = [\dot{p}, \dot{q}, \dot{r}]^T$	deg/s^2	Time derivative of roll, pitch, and yaw rates
$\boldsymbol{\Omega}_b = [p, q, r]^T$	deg/s	Roll, pitch, and yaw rates
$\boldsymbol{\Phi} = [\Phi, \Theta, \Psi]^T$	deg	Roll, pitch, and yaw angles
$\ddot{q}_p$	s^{-2}	Generalized acceleration (structural deflection)
$\dot{q}_p$	s^{-1}	Generalized velocity (structural deflection)
q_p	–	Generalized coordinate (structural deflection)
V_{TAS}	m/s	True airspeed
p_{stat}	Pa	Static atmospheric pressure
$\bar{q}$	Pa	Dynamic pressure
Mach	–	Mach number
α	deg	Angle of attack
β	deg	Angle of sideslip
γ	deg	Angle of climb
χ	deg	Flight-path azimuth
ρ	kg/m^3	Air density
$X_{Aero}, Y_{Aero}, Z_{Aero}$	N	Sum of aerodynamic forces in body frame
$L_{Aero}, M_{Aero}, N_{Aero}$	Nm	Sum of aerodynamic moments in body frame
$X_{TW,l\|r}, Y_{TW,l\|r}, Z_{TW,l\|r}$	N	Forces of left/right engine in body frame
$L_{TW,l\|r}, M_{TW,l\|r}, N_{TW,l\|r}$	Nm	Moments of left/right engine in body frame
$\eta_{l\|r}$	deg	Elevator input, left/right
$\xi_{l\|r}$	deg	Aileron input, left/right
ζ	deg	Rudder input
iH	deg	htp input
$Sp_{i,l\|r}$	Nm	Spoiler (1–5) input, left/right

technologies are currently developed [16–21].[1] A simplified method is to apply the mesh deformation capabilities originally used for elastic structural deformations for the deflection of control surfaces. Therefore, all CFD nodes of the control surface are rotated about a prescribed hinge line and a blending region at the left and right end of the control surface allows for a smooth and continuous transition. Although simple, this method yields reliable and robust results and is used for the time-dependent

[1] In Chap. 1 of this book (Heinrich et al., *Overview of Subproject VitAM-Gust*) new methods and tools for the deflection of complex control surfaces, such as spoilers, are described in detail.

deflection of the primary control surfaces (ailerons and elevators) in the virtual flight tests. The htp is rotated as a whole for the trim and during the maneuver, a sliding approach based on a no-normal movement of the CFD grid nodes of the fuselage adjacent to the htp ensures a valid grid without negative cells in this region as the htp is rotated.

The structural model was generated using an in-house model generator which builds parametric structural models for the MSC. Nastran FE solver based on user-defined geometries (airfoil sections, planform, locations of spars, ribs, and stringers etc.). The model is made from isotropic material (aluminum), a number of sizing loops with mass minimization as the objective function and buckling and thickness constraints leads to an optimized model. The optimization of the model is described in detail in the work of Klimmek and co-authors [13–15]. The global FE model is discretized by several thousand structural elements and nodes. For the virtual flight test–and usually for most kinds of loads analysis–a reduced model is sufficient to represent the dynamics of the structure. To this end, a static condensation (Guyan reduction) is used to obtain a structural model defined by the loads reference axis (LRA) nodes with significantly fewer DOFs.

The full and the condensed finite element models as well as the outer shape of the CFD grid and the unstructured surface mesh of the right wing are illustrated in Fig. 4.

Different methods are available to transfer the aerodynamic loads onto the condensed FE model (it is topologically comparable to a beam model and thus moments must be introduced), such as beam splines, rigid body splines, nearest neighbor mapping, and others. For this work and in general for fluid structure interaction simulations base don CFD aerodynamics and condensed FE models, a couplingmodel approach is used. The couplingmodel is an extension of the condensed structural model by massless, rigid bars. One end of the bars is connected to a particular LRA node, the other is located at the outer surface of the aerodynamic grid. The aerodynamic forces are transferred to the coupling nodes at the outer surface and from there transmitted to the LRA nodes. The couplingmodel is also plotted in Fig. 4.

4 Applications

Selected applications of the high fidelity MDA approach for Virtual Flight Testing are presented in this section. Two longitudinal maneuvers with time dependent inputs of the elevator, aileron, and the horizontal tail plane. The two maneuvers differ in their length, for the second one, the mass properties of the aircraft are modified (reduced moment of inertia about the lateral axis) to increase the dynamics and thus to reduced the simulation time. The third scenario is a 1-cos gust encounter in open loop and with feedforward control for load alleviation. A transfer function was calculated from a time-domain pulse excitation of the aircraft to design the control system. The unsteady maneuver simulations are based on initial conditions which are obtained from a trim simulation that iteratively solves for force and moment

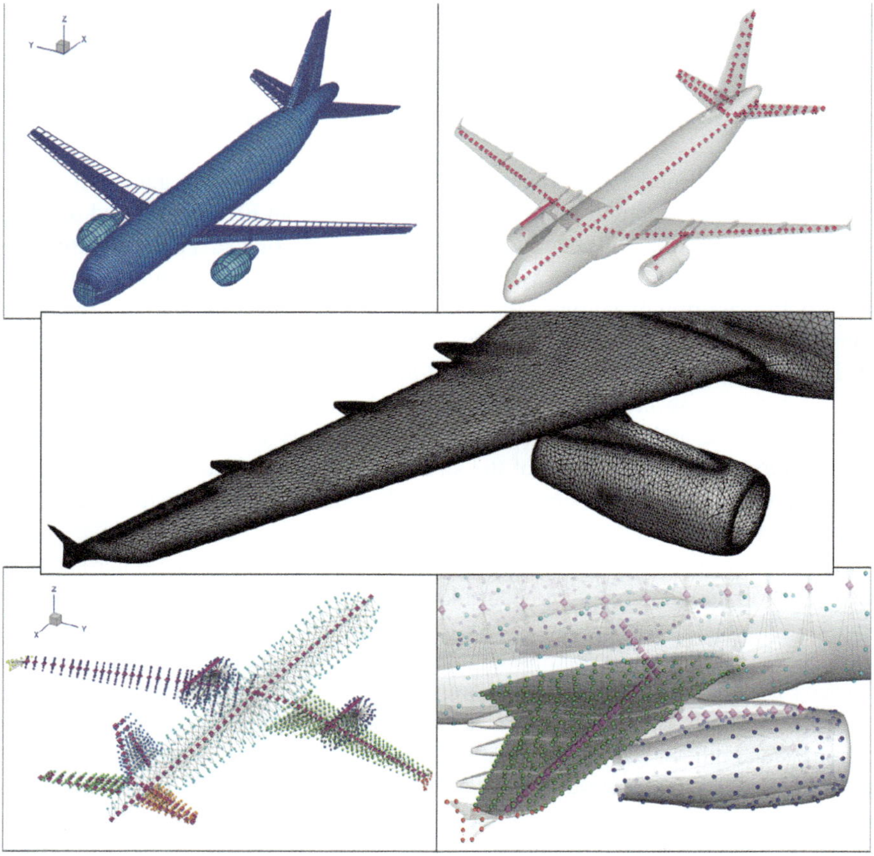

Fig. 4 3D global (blue grid) and condensed (pink dots) finite element model of the jet transport configuration used for the virtual flight tests. Outer shape and CFD surface grid. Engines modeled with in- and outflow engine boundary conditions for realistic thrust generation. The couplingmodel (lower plots) is used to transfer aerodynamic loads onto the nodes of the condensed structural model by massless, rigid bars

equilibrium corresponding to a steady straight level flight as mentioned above. The thrust required to balance the drag forces is modeled directly in the aerodynamic (CFD) model by engine boundary conditions. Based on a table for the particular engine and the airspeed as well as the altitude, the static pressure at the inlet and the total pressure and temperature difference at the outlet are prescribed for the desired thrust setting calculated by the trim solver. From the trim simulation, the AoA (the pitch angle), the angle of the horizontal tail plane, the thrust of the engines, and the structural deformations (in terms of the generalized coordinates q_k) are obtained. The parameters (aircraft with particular mass model, atmospheric data, gust field) are based on the *VitAM-Gust* scenario mentioned in the introduction and listed in Table 2.

Table 2 Parameters specified by the *VitAM-Gust* project for the virtual flight tests presented in this section

Parameter	Unit	*Virtual flight tests DLR-ATRA*
Aircraft mass	kg	58200
Mach number	–	0.78
Altitude	ft	37000
Airspeed V_{TAS} (trim)	m/s	230.13
MAC	m	4.1935
Reynolds number (MAC)	–	$23.64e^6$
Density	kg/m^3	0.3484
Static temperature	K	216.65
Static pressure	kPa	21.6627
Additional parameters for GLA scenario		
Gust length	m	213.36
Gust velocity	m/s	10.52

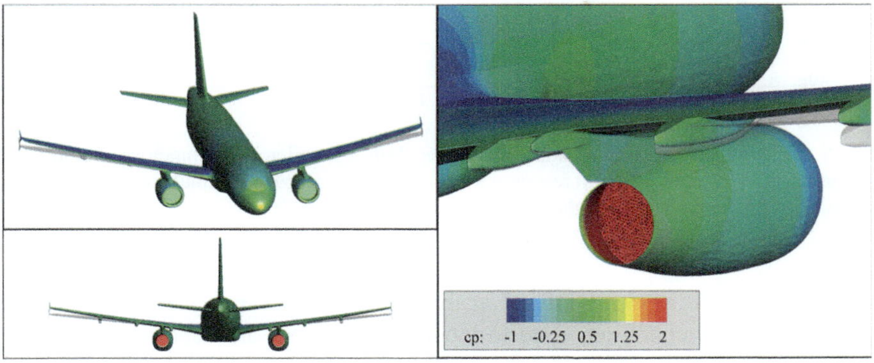

Fig. 5 Trim simulation of the jet transport with engine thrust modeled by engine boundary condition in the CFD solver

The structural deformation at steady straight and level flight is depicted exemplarily in Fig. 5 together with a color map illustrating the pressure distribution.

4.1 *Longitudinal Maneuver for System Identification*

Exemplary results of a longitudinal system identification maneuver (virtual system identification flight) with inputs for the horizontal tail plane, the elevator, and the ailerons (symmetric) are presented first. For the unsteady simulations, the physical timestepsize is 0.005 s and the flight dynamics equations of motion are integrated

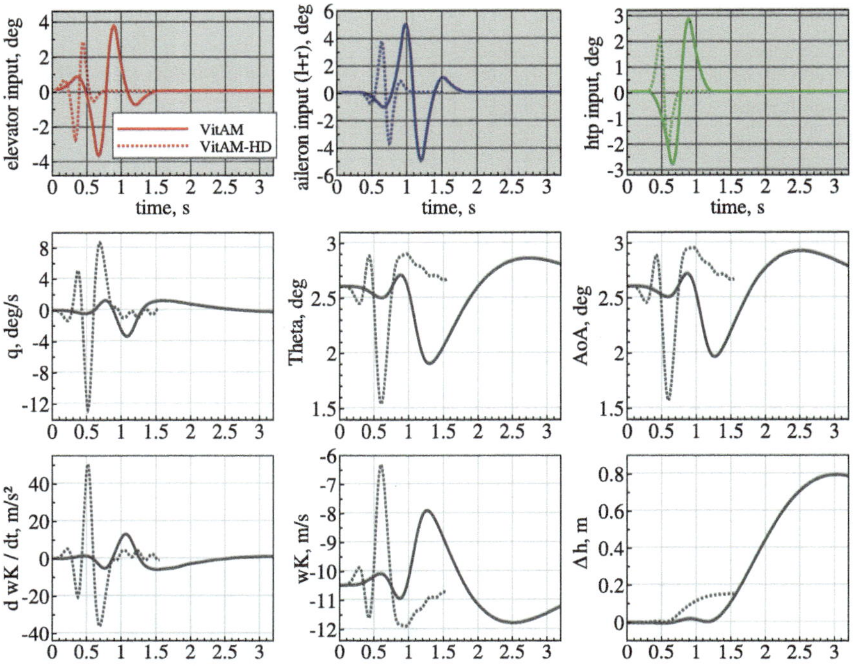

Fig. 6 Virtual flight test of longitudinal maneuver: Inputs for ailerons (symmetric), elevator, and horizontal tail plane (iH); Aircraft response for normal inertia properties (simulation time 3.2 s) and for reduced pitch inertia (HD, higher dynamics, simulation time 1.6 s)

in time by a semi-implicit ODE solver from the SciPy library[2]. The settings for the engine boundary conditions used to model thrust forces are kept constant during the entire maneuver, unsteady inputs for the control surfaces are prescribed for each time step. For the identification of the aircraft's longitudinal characteristics–which is the goal of the activities of the project *VitAM-Gust* –, only inputs which excite the longitudinal motion are used. Two kinds of inputs with different total flight times were developed. The standard signals enable the identification from a flight time of 3.2 s. By modification of particular parameters of the aircraft with a first order effect on the dynamical behavior, the total flight time can be reduced. Here, the moment of inertia about the lateral axis, I_{yy}, was divided by a factor of ten to significantly increase the pitch dynamics. Using this approach, the flight time could be reduced to 1.6 s. Inputs and results of the longitudinal maneuver are plotted in Fig. 6. The wavelets, which were chosen as control surface inputs for this virtual flight test, can be seen in all control surface inputs. Significantly increased dynamic responses of the aircraft in terms of the longitudinal states are obtained from the simulation with the reduced inertia properties. High frequencies in the pitch rate and in the

[2] https://scipy.org/.

acceleration in the z direction appear, which, among others, have their origin in the stronger excitation of the elastic structural modes. Those results–together with various other flight mechanic state vectors and the generalized coordinates–are used for the subsequent system identification process described in Chap. 5 of this book (Roeser and Mönnich, *System Identification from Virtual Flight Test Data for a Flexible Transport Aircraft*).

4.1.1 1-Cos Gust Encounter Scenario with Feedforward Gust Load Alleviation

The third application of the MDA framework for virtual flight testing is a 1-cos gust encounter maneuver, as it is demanded by certification regulations, such as the European CS 25, *Certification Specifications for Large Aeroplanes*, (cf. Book 2, Subpart C: *Gust and Continuous Turbulence Design Criteria*) [22]. This bookcase scenario is already used during the preliminary design phase of an aircraft because–especially for large jet transport aircraft with slender wings–gust loads are typically part of the set of design load cases. The gust loads–especially the bending moments–can be reduced by gust load alleviation (GLA) methods, reducing their contribution to the loads that must be considered for the structural sizing [23]. As the modeling of dynamic spoiler deflections in the CFD solver is currently in process, the load alleviation will be achieved by symmetric deflection of the ailerons to relieve the lift at the outer sections of the wing and thus to reduce the bending moment. Since the largest gust loads typically emerge at frequencies near the first wing bending mode–which is higher than the short period mode–and elevators are seen as effective only up to the frequency of the short period mode (which is below 1 Hz), the ailerons are used for the realization of the GLA system.

A careful design of the GLA system is required, however, and the transfer function (TF) with the symmetric aileron deflection as input and the structural loads as output (e.g. cut or nodal forces and moments along the wing) must be determined with high quality. To this end, we present a simple, yet effective approach for the calculation of such a transfer function. A time-domain simulation (physical time step size 0.005 s) of the aircraft in free-flight with a specially designed pulse function that excites the airframe by a symmetric deflection of the aileron is used to identify this transfer function. Therefore, the time domain results of the structural loads (shear force and bending moment about the x axis at the wing root), calculated by the force summation method, and the excitation signal are transformed into the frequency domain. The pulse function in the time domain and its frequency content are depicted in Fig. 7.

The rigid-body and elastic responses of the aircraft were transformed into the frequency domain by means of an FFT algorithm to calculate the transfer function, which is plotted in Fig. 8. The plot shows dominant impacts of the (symmetric) elastic structural modes. Using this TF the effectiveness of the GLA in reducing the loads as a function of frequency is obtained and considered in the GLA synthesis. For simplicity and because the focus is on the application of the MDA framework in

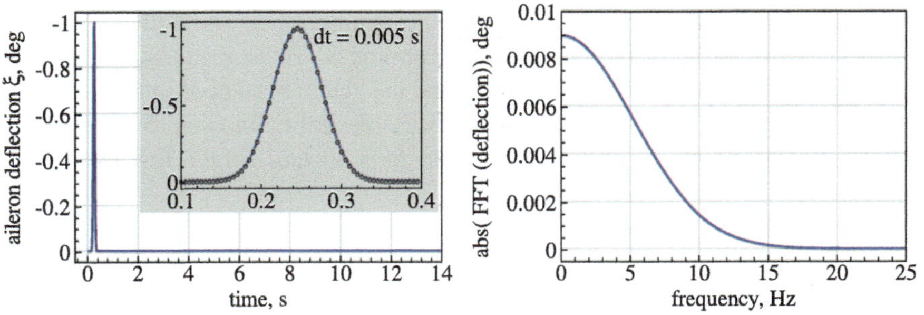

Fig. 7 Time history and frequency content of the pulse function used for the excitation of the aircraft in free-flight by (symmetric) deflection of the ailerons

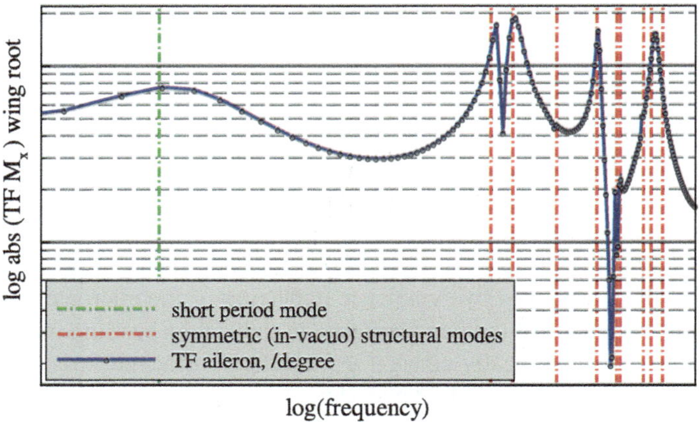

Fig. 8 Transfer function of jet transport aircraft relating aileron deflection (per degree) to wing root bending moment M_x. The model was linearized about the steady trim state in free-flight

this work, only basic control system elements are used: Proportional terms as well as low-pass and high-pass filters) Proper time-delays adjusted to the airspeed were used such that the ailerons are deflected after a constant distance of gust penetration to prevent premature or delayed deflecting of the ailerons. Nonlinearities are included to limit the rate of the aileron deflection to 40°/s (this limiter only intervenes at short and medium gust lengths). An angle of attack sensor positioned near the nose of the aircraft is used to measure of the disturbance (gust) which serves as input to the GLA system. The function of the feedforward GLA is depicted in the block diagram of Fig. 9. A 1-cos gust encounter scenario was selected to demonstrate the impact of the feedforward GLA system. As before, the physical timestep size of the simulation is 0.005 s. Results of the flight dynamics simulation of the gust encounter are presented in Fig. 10 in terms of the rigid-body pitch rate and -acceleration, the load factor in the z direction, as well as the bending moment about the x axis at the wing root. It must be mentioned that the feedforward control applied for this test case is not optimized (neither the gains, the delays, or the filters have been adjusted

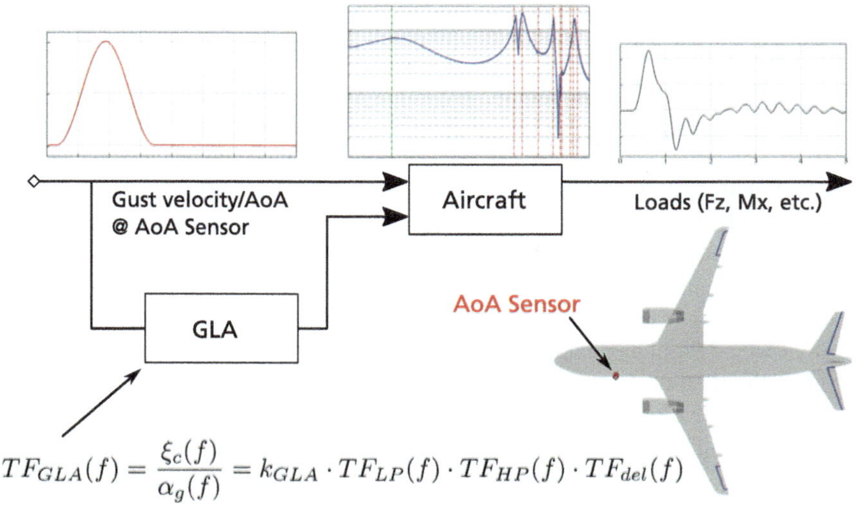

Fig. 9 Illustration of the feedforward GLA system

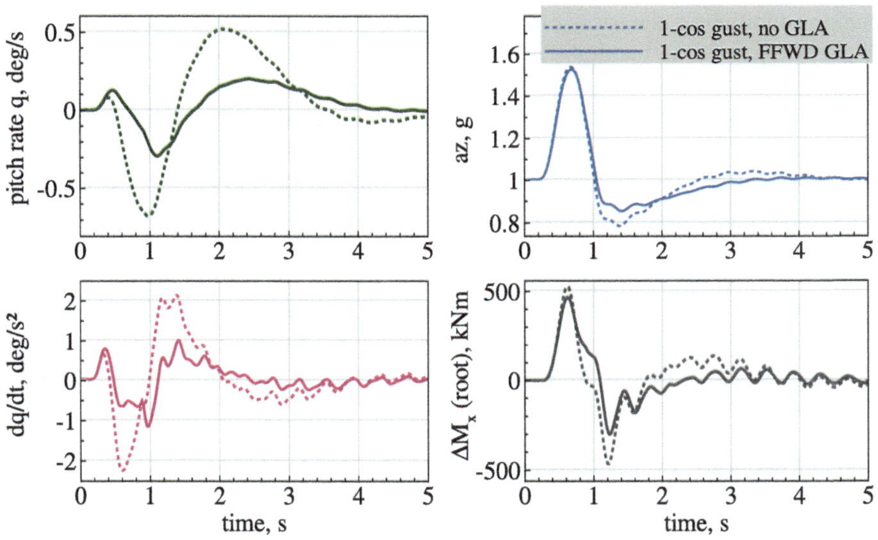

Fig. 10 1-cos gust encounter scenario: Rigid-body states and bending moment at wing root for open loop and feedforward GLA simulation of jet transport aircraft

to yield optimal performance), the intended purpose of this application is simply the demonstration of how such a control system can be designed based entirely on MDA simulations. Such simulations can be performed as soon as the necessary simulation models are available in the (preliminary) design stage of the aircraft.

5 Summary and Outlook

This paper presents a high fidelity, multidisciplinary simulation approach for the virtual flight testing to determine an aircraft's longitudinal flight mechanic characteristics. The approach uses very accurate simulation models for the aerodynamics (URANS with a CFD grid build from a very detailed CAD model) and the structural dynamics (FE model in loads reference axis representation). All disciplines are coupled in space and time. For the aircraft identification process, which uses selected states of the flight dynamics simulation, carefully designed input signals for the elevator, the ailerons (symmetric), and the horizontal tail plane with frequencies close to the expected flight mechanic eigenvalues are applied. The aircraft can be identified with this longitudinal input from a flight time of only 3.2 s. By reducing the inertial properties of the aircraft about the lateral axis and thus increasing the longitudinal dynamics, a flight time of 1.6 s is sufficient. Another virtual flight test is presented in terms of a 1-cos gust encounter scenario with a feedforward load alleviation controller, where a transfer function, identified from the time domain results of a high fidelity simulation with the same model, was used for the design of the controller.

The applications and simulations presented in this work are characterized by rather benign conditions, i.e. unsteady maneuver simulations about a steady trim point close to the design point in the flight envelope. The complexity of the analyses is increasing considerably, however, if flight points away from the nominal operation region would be considered. Nonlinearities mainly in the aerodynamics require modeling technologies that surpass the methods applied until now (so far, only URANS was used). In addition, the configuration of the aircraft might be much more complex if, for instance, low speed conditions with deployed high lift devices are investigated. Issues regarding the convergence and the robustness of the solution must be expected (apart from that, the modeling and computational costs are further increasing). The complexity will increase on the structural side as well, as more detailed models are used and nonlinearities from geometrically large deflections need to be considered in future. As can be seen from these examples, a great deal of extensions is necessary in terms of physical modeling capabilities.

Acknowledgements The funding of partial activities through the Federal Ministry for Economic Affairs and Climate Action under contracts FKZ 20A1504A in the framework of the air transport research program is gratefully acknowledged.

References

1. Roeser, M.S.: Multi-Axis Maneuver Design for Aircraft Parameter Estimation, Deutscher Luft- und Raumfahrtkongress (DLRK). Deutsche Gesellschaft für Luft- und Raumfahrt - Lilienthal-Oberth e.V, Friedrichshafen, Germany (2018)
2. Waszak, M.R., Schmidt, D.K.: Flight Dynamics of aeroelastic vehicles. J. Aircr. **25**(6), (1988). American Institute of Aeronautics and Astronautics

3. Buttrill, C., Arbuckle, P., Zeiler, T.: Nonlinear Simulation of a Flexible Aircraft in Maneuvering Flight. In: Guidance, Navigation, and Control and Co-located Conferences. American Institute of Aeronautics and Astronautics, Monterey, California (1987)
4. Meirovitch, L.: Hybrid state equations of motion for flexible bodies in terms of quasi-coordinates. J. Guid. Control. Dyn. **14**(5), (1991). American Institute of Aeronautics and Astronautics
5. Ritter, M., Jones, J., Cesnik, C.E.S.: Free-flight Nonlinear Aeroelastic Simulations of the X-HALE UAV by an Extended Modal Approach. In: International Forum on Aeroelasticity and Structural Dynamics, IFASD, Como, Italy (2017)
6. Beckert, A., Wendland, H.: Multivariate interpolation for fluid-structure-interaction problems using radial basis functions. Aerosp. Sci. Technol. **5**(2), (2001). Elsevier
7. ZONA Technology Inc., ZAERO Theoretical Manual Version 8.4 (2011)
8. Schwamborn, D., Gerhold, T., Heinrich, R.: The DLR TAU-code: Recent applications in research and industry. In: ECCOMAS CFD 2006: Proceedings of the European Conference on Computational Fluid Dynamics, Egmond aan Zee, The Netherlands (2006)
9. Meinel, M., Einarsson, G.: The FlowSimulator framework for massively parallel CFD applications. In: PARA 2010. Reykjavik, Island (2010)
10. Reimer, L., Ritter, M., Heinrich, R., Krüger, W.: CFD-based gust load analysis for a free-flying flexible passenger aircraft in comparison to a DLM-based approach. In: 22nd AIAA Computational Fluid Dynamics Conference. American Institute of Aeronautics and Astronautics, Dallas, Texas (2015)
11. Reimer, L., Heinrich, R., Ritter, M.: Progress in the multidisciplinary simulation environment FlowSimulator towards a virtual flight testing capability of aircraft. In: DGLR/STAB Symposium, Jahresbericht 2018. Darmstadt, Germany (2018)
12. Ritter, M., Roeser, M.S., Reimer, L.: CFD-based multi-axis maneuver simulation for system identification of flexible transport aircraft. In: AIAA SciTech 2020 Forum. Orlando, USA (2020)
13. Klimmek, T.: Parameterization of topology and geometry for the multidisciplinary optimization of wing structures. In: Proceedings of CEAS European Air and Space Conference (CEAS2009), Manchester, UK (2009)
14. Klimmek, T.: Parametric set-up of a structural model for FERMAT configuration for aeroelastic and loads analysis. J. Aeroelasticity Struct. Dyn. **3**(2), (2014)
15. Bramsiepe, K.R.B., et al.: Loads and structural optimisation process for composite long range transport aircraft configuration. In: 19th AIAA/ISSMO Multidisciplinary Analysis and Optimization Conference, Atlanta, GA (2018)
16. Spiering, F.: Development of a fully automatic chimera hole cutting procedure in the DLR TAU code. In: New Results in Numerical and Experimental Fluid Mechanics X, vol. 132. Springer International Publishing Switzerland (2016)
17. Reimer, L., Heinrich, R.: Modeling of movable control surfaces and atmospheric effects. In: Notes on Numerical Fluid Mechanics and Multidisciplinary Design, Springer Series Notes on Numerical Fluid Mechanics and Multidisciplinary Design (Computational Flight Testing), vol. 123 (2013)
18. Reimer, L., Heinrich, R., Ritter, M.: Progress in the multidisciplinary simulation environment FlowSimulator towards a virtual flight testing capability of aircraft. In: Jahresbericht 2018, DGLR/STAB Symposium, Darmstadt, Germany (2018)
19. Capsada, L.A., Heinrich, R.: Development of the DLR TAU code for modelling of control surfaces, Deutscher Luft- und Raumfahrtkongress (DLRK). Deutsche Gesellschaft für Luft- und Raumfahrt - Lilienthal-Oberth e.V, Darmstadt, Germany (2019)
20. Geisbauer, S., Löser, T.: Towards the investigation of unsteady spoiler aerodynamics. In: 35th AIAA Applied Aerodynamics Conference, Denver, Colorado (2017)
21. Geisbauer, S.: Numerical Simulation and Validation of the Aerodynamics of Static and Dynamic Spoilers, Deutscher Luft- und Raumfahrtkongress (DLRK). Deutsche Gesellschaft für Luft- und Raumfahrt - Lilienthal-Oberth e.V, Friedrichshafen, Germany (2018)

22. Agency, E.A.S.: Certification specifications for large aeroplanes CS-25. Amendment **5**, 5 (2008). September
23. Handojo, V., Lancelot, P., De Breuker, R.: Implementation of active and passive loads alleviation methods on a generic mid-range aircraft configuration. In: 2018 Aviation Technology, Integration, and Operations Conference, Atlanta, Georgia (2018)
24. Brockhaus, R.: Flugregelung. Springer, Berlin Heidelberg (2013)

Robustness Improvements for a Differential Reynolds Stress Modeling Using a Logarithmic Formulation

Sebastian Braun and Vamshi Togiti

Abstract A new version of the SSG/LRR Reynolds stress model that uses the natural logarithm of the specific dissipation rate ω as the length-scale providing variable is developed within the Vitam project and implemented into the DLR-TAU code. A verification and validation of the model implementation are conducted using several test cases which covered wall bounded flows as well as separating flows. Performance assessment of the new version of the model is carried out by comparing the predictions with SSG/LRR-ω predictions and experimental data for complex industry relevant configurations. Comparison studies show that the new version of the model improves the robustness and convergence of computations while delivering predictions similar to the SSG/LRR-ω predictions.

Keywords Robustness of turbulence models · Reynolds-stress model · Length-scale-providing variable

1 Introduction

Significant enhancements in computational fluid dynamics (CFD) have led to the use of numerical simulation in aircraft design process. The CFD methods used in the design process rely on the Reynold-averaged Navier–Stokes equations and some type of turbulence modelling in most cases. Often Eddy viscosity models (EVM) are employed to model turbulence. However, the accuracy of EVM, like the Spalart–Allmaras (SA) model [1] or the shear-stress transport model (SST) published by Menter [2], seem to be insufficient at the borders of the flight envelope where massive flow separation dominates flight physics. On the contrary, dropping the Boussinesq hypothesis used in EVMs and modelling the transport equations for the Reynolds stresses directly allows for a better description of flow physics. Though,

S. Braun · V. Togiti (✉)
Center for Computer Applications in AeroSpace Science and Engineering ($C^2A^2S^2E$),
Institute of Aerodynamics and Flow Technology, Braunschweig, Germany
e-mail: vamshi.togiti@dlr.de

R. Heinrich (ed.), *Advanced Aircraft Understanding via the Virtual Aircraft Model*,
Notes on Numerical Fluid Mechanics and Multidisciplinary Design 155,
https://doi.org/10.1007/978-3-031-69425-7_3

these Reynolds stress models (RSM) are often reported to be hard to solve and to degenerate the robustness and stability of the solution process in comparison to state of the art EVMs. Nevertheless, the SSG/LRR-ω-RSM developed by Eisfeld [3] has been demonstrated to be applicable to a wide range of aeronautical flow problems [4] and is now routinely used within DLR and industry for fairly complex cases. However, stability and robustness issues remain a problem especially in more complex cases like high-lift configurations for example.

In order to improve the robustness of RSM, a new formulation of the SSG/LRR-RSM is presented in which the specific dissipation rate ω is replaced by its natural logarithm. This transformation is based on ideas of Bassi et al. [7] that were developed in the framework of two-equation turbulence model. Due to this transformation, extremely large slope of ω near viscous wall is alleviated and the necessity of a numerical limitation of the variable in the event of negative length-scale variables is eliminated and thus the robustness and stability of the RSM is improved.

In this paper, the verification, validation and application of the ln(ω) length-scale equation in conjunction with the SSG/LRR-RSM, here on denoted as RSM-ln(ω), is presented. In the verification study, the correctness of the model formulation and its implementation into the TAU code is examined. In the validation studies, 2D separated flow test cases are investigated and the predictions are compared to the experimental data. In the applicability study, investigations are conducted on industry relevant configurations and the performance of the new formulation of the RSM-ln(ω) is assessed.

2 Turbulence Model Equations

The present work is based on the SSG/LRR-ω Reynolds stress model [3, 4] developed at DLR. This model uses the Menter Baseline ω-equation [2] for providing the length-scale. To improve the robustness of the SSG/LRR-RSM, Braun transformed the ω-equation into $ln(\omega)$-equation by recasting $\omega = e^{ln(\omega)}$ in the Menter baseline ω-equation [2] and implemented into the DLR-TAU code.

The transport equation for $ln(\omega)$ is given in the following:

$$\tilde{\omega} = ln(\omega) \tag{1}$$

$$\begin{aligned}\frac{\partial\left(\overline{\rho}\tilde{\omega}\right)}{\partial t} + \frac{\partial\left(\overline{\rho}\tilde{\omega}U_k\right)}{\partial x_k} = {} & \frac{\alpha_{\tilde{\omega}}}{k}\frac{\overline{\rho}P_{kk}}{2} - \beta\overline{\rho}e^{\tilde{\omega}} + \frac{\partial}{\partial x_k}\left[\left(\mu + \sigma_{\tilde{\omega}}\frac{\rho k}{e^{\tilde{\omega}}}\right)\frac{\partial\tilde{\omega}}{\partial x_k}\right] \\ & + \left(\mu + \sigma_{\tilde{\omega}}\frac{\rho k}{e^{\tilde{\omega}}}\right)\frac{\partial\tilde{\omega}}{\partial x_k}\frac{\partial\tilde{\omega}}{\partial x_k} \\ & + \sigma_d\frac{\rho}{e^{\tilde{\omega}}}\max\left(\frac{\partial k}{\partial x_k}\frac{\partial\tilde{\omega}}{\partial x_k}, 0\right)\end{aligned} \tag{2}$$

The transformed $\tilde{\omega}$ transport equation is coupled to the SSG/LRR-RSM through the isotropic dissipation rate ϵ that is computed as follows:

$$\epsilon = C_\mu k \omega \tag{3}$$

The transport equations of the SSG/LRR model remain unchanged. Complete details of the SSG/LRR-RSM model coupled to $\ln(\omega)$ transport equation can be found in [5, 6].

3 Results

In this section, the verification, validation and application of the RSM-$\ln(\omega)$ are discussed.

3.1 NACA0012 Verification Study

In order to investigate the effects of the spatial discretization in more detail, a mesh refinement study was conducted using the well-known NACA0012 airfoil [8]. For this test case, large number of meshes are available with resolutions ranging from a few thousand grid points up to several million points. As the RSM-$\ln(\omega)$ model is obtained by a purely mathematical transformation of the RSM-(ω) model, both versions are expected to deliver identical results in the limit of infinitely refined meshes. The flow condition used in this study are a Mach number of $M = 0.15$ and a chord based Reynolds number of 6 million. The angle of attack was set to 10°. Fig. 1 shows the lift coefficient as well as the drag coefficient for different spatial mesh resolutions $h = 1/\sqrt{N}$, where N represents the number of grid points of the corresponding mesh. Starting with an initial mesh that consists of barely 29,152 grid points, the number of points is quadrupled in each refinement step which leads to a final resolution of 29.4 million grid points after 5 refinement steps. The lift and drag coefficients computed on these six meshes are plotted in Fig. 1. While large deviations of the generated lift and drag can be found on the coarse meshes, these deviations vanish with an increasing number of grid points and becoming nearly incognizable on the finest mesh as expected for a correctly implemented model.

3.2 Backward Facing Step Validation Test Case (2D)

In this case, the oncoming flow separates due to sudden change in the geometry, back step, then reattaches and recovers downstream of the back step. In this study, computations are carried out at a Reynolds number based on the height of the back

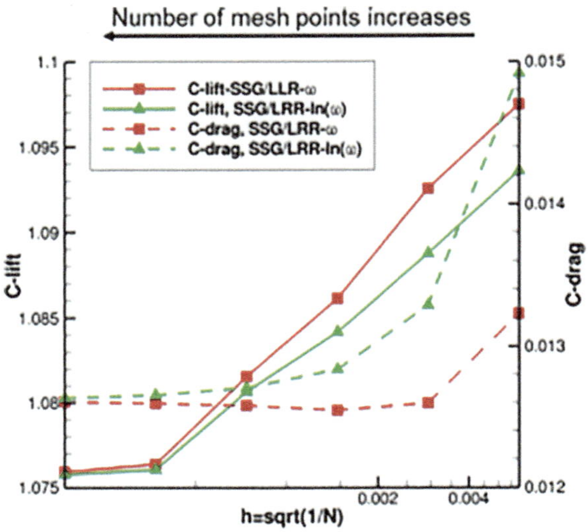

Fig. 1 Mesh refinement study for lift and drag coefficients using the NACA0012 airfoil. Meshes are provided by NASA on the TMR-Website [9]. The mesh resolution is ranging from $N = 29,152$ grid points to $N = 29.4$ million points

step $Re_H = 36000$ and Mach number $M = 0.128$ on the second finest grid available on the NASA TMR website [9].

In Figs. 2 and 3, pressure- and skin-friction coefficient distribution along the back-step wall and velocity and Reynolds shear stress profiles downstream of the back-step delivered by RSM-ln(ω) are compared to the RSM-(ω) results. It is apparent from the figures that the c_p, c_f, velocity and Reynolds shear stresses predicted by the different variants are indistinguishable.

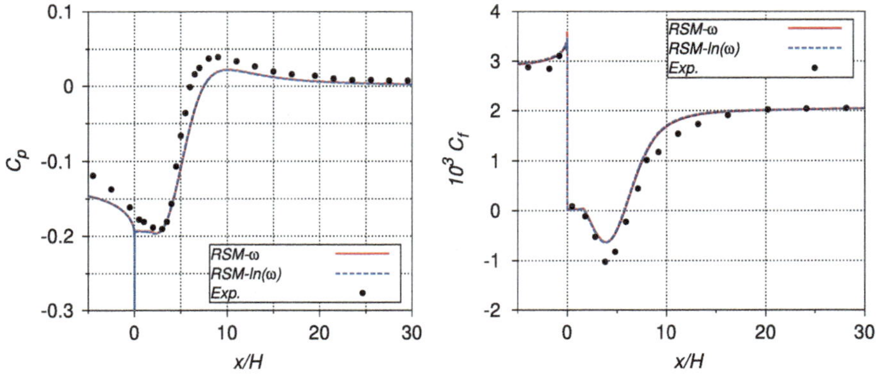

Fig. 2 Comparison of different variants of RSM for the backward facing step. Left: c_p, right: c_f

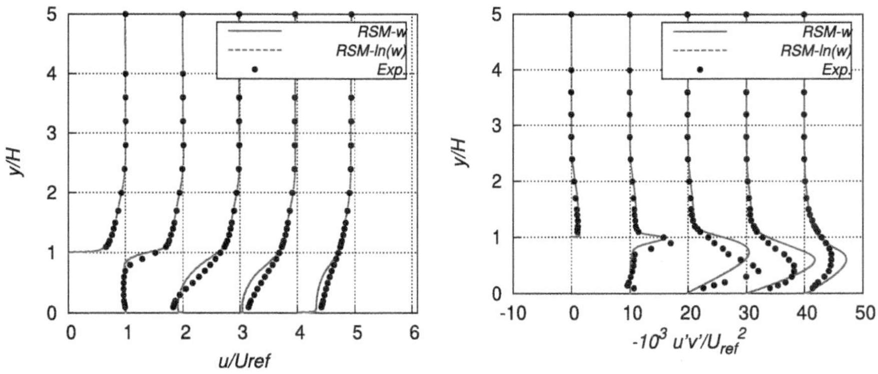

Fig. 3 Comparison of different variants of RSM for the backward facing step. Left: velocity profile, right: Reynolds shear stress profile

3.3 NASA Wall-Mounted Hump Validation Test Case (2D)

In the wall-mounted hump case, the oncoming flow separates on the hump due the adverse pressure gradient imposed by downstream [10]. For this case, computations are carried out at a Reynolds number $Re_c = 936000$ and Mach number $M = 0.1$. The computational grid consisting of 817×217 points available on the NASA TMR website [9] is used.

Convergence histories of the residuals of density, Reynolds shear stresses and length-scale are shown in Fig. 4. For all the residuals, the RSM-ln(ω) delivers improved convergence compared to the RSM-(ω).

In Fig. 5, pressure- and skin-friction coefficient along the hump wall is shown. For c_p , all three variants show differences in the separated region ($0.6 < x/c < 1.2$). The reason for differences is the size of separation. To vizualize separation extent, c_f along the hump wall is shown in Fig. 5 (right). The region with negative c_f indicated the extent of flow separation. As can been in the figure, the separation extent is smaller in RSM-ln(ω) and RSM-g [11] compared to RSM-(ω). Since ln(ω) and g-equations are obtained by transforming the ω equation, it is anticipated that the results would be identical on this fine grid. However, for this flow, the grid chosen may be still not fine enough to reduce numerical errors introduced in the additional gradient terms which arose from transformation (ω to $\ln(\omega)/g$). To verify this, computations are carried out on the finest grid consisting of 1633×433 points.

In Fig. 6, comparison of c_p and c_f is shown. On the finest grid almost similar pressure distribution is predicted. However for c_f , still some difference is observed for the reattachment point. It may be that computations on even finer grids are need to obtain grid converged results for RSM-ln(ω) and RSM-g.

Overall, for this wall-mounted hump cases, the RSM-ln(ω) improves the convergence compared to the other formulations.

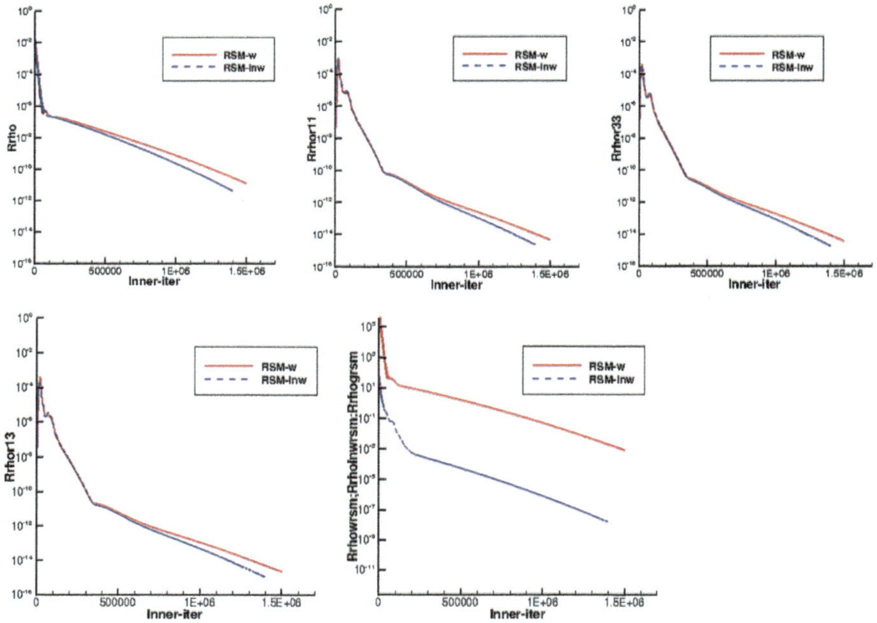

Fig. 4 Convergence history for the wall-mounted hump on the second finest grid, 817 × 217

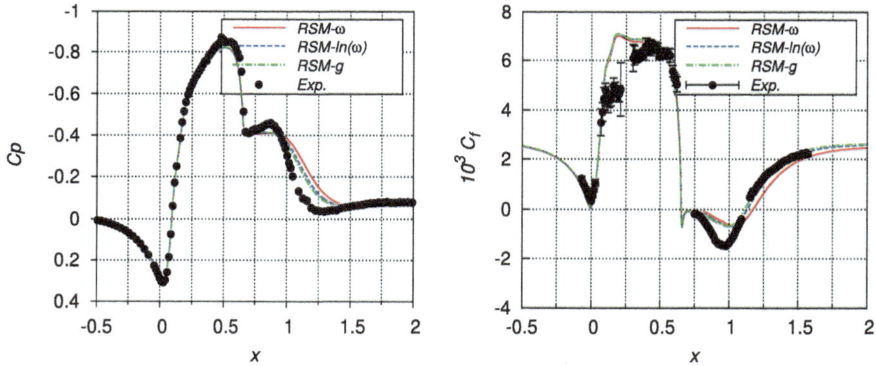

Fig. 5 Comparison of different variants of RSM for the NASA wall-mounted hump on the second finest grid, 817 × 217. Left: c_p, right: c_f

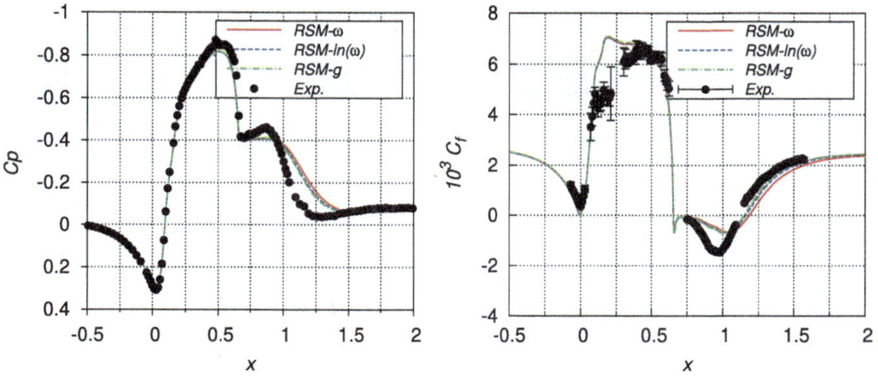

Fig. 6 Comparison of different variants of RSM for the NASA wall-mounted hump on the finest grid, 1633 × 433. Left: c_p, right: c_f

4 Applications to Industrial Flows

In this section, the applicability of RSM-ln(ω) for industrial flows is discussed.

4.1 CRM

The NASA Common Research Model (CRM) [12], see Fig. 7, is used to perform stability and convergence tests with the SSG/LRR-RSM.

Here, comparative computation tests with the RSM-ω, RSM-g, RSM-ln(ω) at high CFL numbers are performed. The used mesh was provided by NASA in the framework of the sixth AIAA drag prediction workshop [14]. Prisms are used within

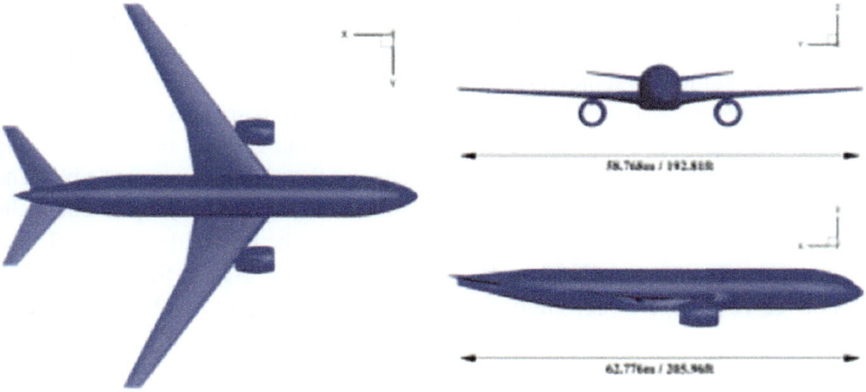

Fig. 7 Scaled CAD model of the NASA common research model. Adopted from [13]

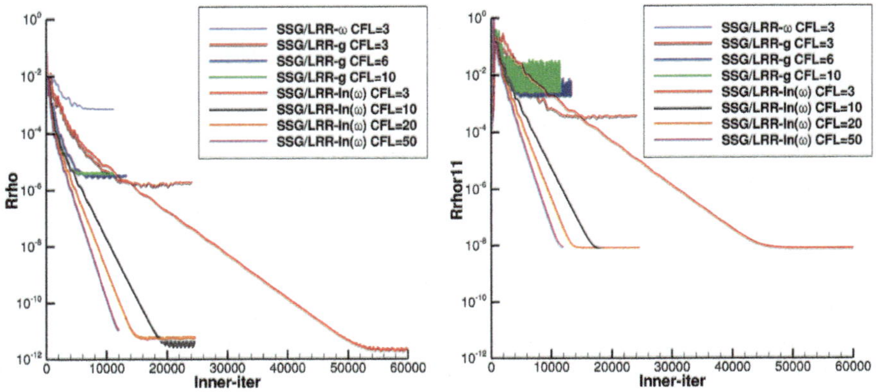

Fig. 8 Convergence history for the NASA-CRM at $\alpha = 4°$ and $Ma = 0.85$ computed with three different versions of the SSG/LRR-RSM

this mesh in order to discretize the flow near viscous walls and tetrahedrons are used everywhere else. In total, the mesh contains 42.8 million grid points. The simulations were conducted at an angle of attack of $\alpha = 4°$ and a Mach number of $M = 0.85$. Several simulations using a three stage multigrid-approach were performed under these conditions with CFL-numbers in the range of 3–50. The time integration was performed by using the implicit Backward–Euler scheme. The resulting convergence histories for the residuals of density and the Reynolds stress component R11 are shown in Fig. 8. When using the RSM-(ω) model in combination with a CFL-number of 3, the density residual is reduced by merely 3 orders of magnitude before the solution procedure stalls. Deactivating the multigrid algorithm or a further reduction of the CFL number did not improve this. After switching from the RSM-(ω) formulation to the SSG/LRR-g-model, a much larger reduction in the density residual of more than 5 orders of magnitude can be observed. The residuals of Reynolds stresses are reduced by 3 orders of magnitude in case of the g-model-version but less than one order of magnitude when the original RSM-(ω) version is used. In order to check the stability limits of the SSG/LRR-g model, the CFL-number was increased in two steps from 3 to values of 6 and 10. Increasing the CFL number divides the number of iterations required to achieve the same reduction of the density residuals in halves. However, large oscillations of the residuals of the Reynolds stresses can be observed and the achievable convergence level is reduced to values bigger than 10^{-2}. Using the $\ln(\omega)$-formulation improves the convergence of the simulation remarkably. The density residual is reduced by nearly twelve orders of magnitude independent of the used CFL number. The same holds for the residuals of the Reynolds stresses. There, a reduction to values smaller than 10^{-8} can be achieved. Again, the CFL number was increased in several steps in order to check the stability limits. No convergence issues or stability problems can be observed up to a CFL number of 50. Using such high CFL numbers reduces the cost of the simulation by about 70% as the number of required iterations decreases from more than 50.000 to less than 14.000.

4.2 DLR-F11

The configuration studied in the High-lift prediction workshop 2 (HiLiftPW-2) [15] is investigated in the present study. First, the high-lift configuration without slat tracks and flap track fairings is investigated. Later, polar computations are carried out for the high-lift configurations with slat tracks and flap track fairings. All the investigations are carried out on the medium grid (Grid D) available on the HiLiftPW-2 website [15].

For the high-lift configuration without slat-tracks and flap-track fairings, computations are carried out at a Reynolds number of 15×10^6 and Mach number of 0.175 and the incidence angle of 16° with RSM-(ω) and RSM-ln(ω). The objective of this study is to demonstrate the advantage of RSM-ln(ω) over the RSM-(ω) for complex industrial flows. Both RSM variants show convergence issues when 3v multigrid cycle is used. Because of this, computations are run with a 2v multigrid (MG) cycle initially, and later computations are carried out with a singlegrid (SG) to reduce oscillations in residuals and force coefficients.

The convergence histories for the residuals of density, Reynolds stress and length-scale variables are shown in Fig. 9. It is evident from the figure that with the RSM-ln(ω) better convergence is achieved compared to RSM-(ω).

Concerning oscillations in residuals, when singlegrid is used oscillations are reduced with the RSM-ln(ω) while with the RSM-(ω) no change in the residual oscillations is observed. Concerning lift, both RSM variants delivered same lift coefficient. Better convergence and same lift coefficient show the advantage of ln(ω)-equation over the ω-equation in conjunction with the RSM for complex high-lift configurations.

For the high-lift configuration with slat-tracks and flap-track fairings, computations with RSM-(ω) crashed. However, with the RSM-ln(ω) such a trend is not observed for the complete polar. For comparison, the RSM-g results, obtained in earlier studies which were observed to be almost close to the RSM-(ω) predictions delivered by FUN3D code, are used.

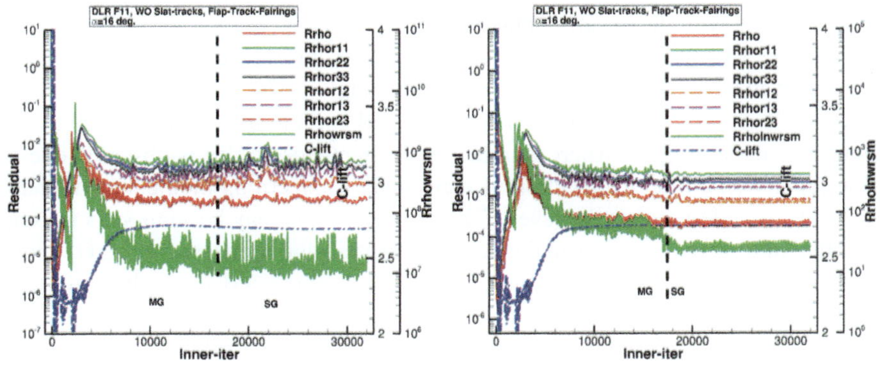

Fig. 9 Convergence history for the configuration without slat-tracks and flap-track fairings at $\alpha = 16°$. Left: RSM-(ω), right: RSM-ln(ω)

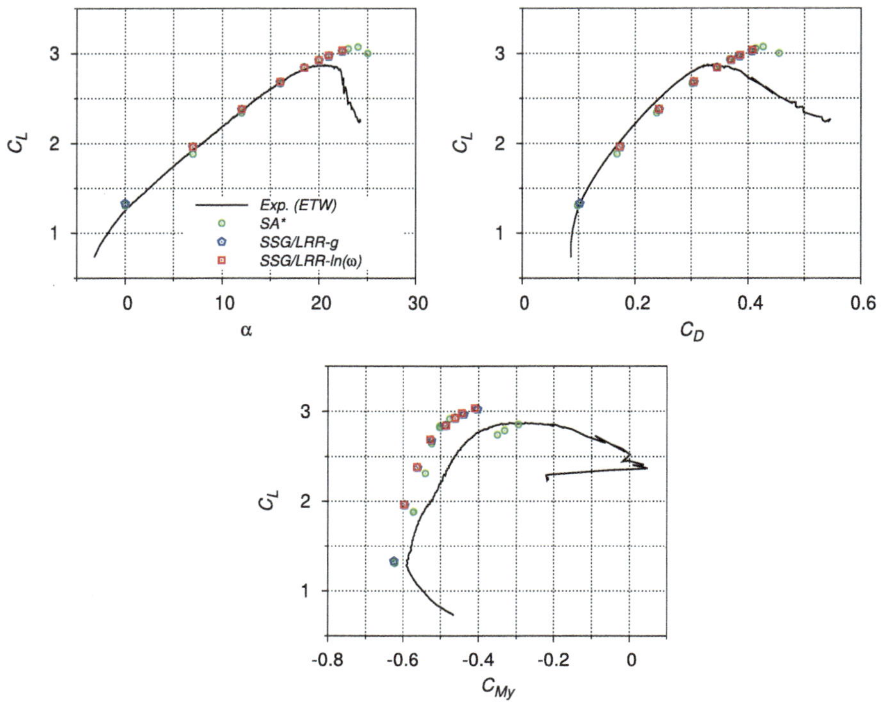

Fig. 10 Force coefficients for the DLR F11. Left: lift, center: drag, right: Moment

Polar computations are run starting from the incidence angle of $\alpha = 7°$ to 22.5° with $\Delta\alpha = 1°$. Concerning convergence, for all the computations RSM-ln(ω) delivered convergence similar to the convergence shown for the case without slat tracks and flap track fairings. With RSM-g convergence similar to RSM-(ω), see Fig. 9 (left), was observed.

Force and moment coefficients delivered by RSM-ln(ω) and RSM-g are shown and compared with the experimental data in Fig. 10.

Here the SA results are shown as reference. RSM-ln(ω) results show good agreement with the experiments up to the incidence angle of 20° for lift and drag coefficients. Beyond $\alpha = 20°$, higher lift is delivered and the trend continues until the incidence angle of 22.5° whereas in the experiments loss of lift is observed due to flow separation on the mid section of main wing. Concerning moment coefficient, RSM-ln(ω) underpredicts the levels. The reason for this may be semi-span testing, as stated in Ref. [16]. Similar to the verification and validation test cases, very close agreement between the predictions of RSM-ln(ω) and RSM-g for the force and moment coefficients is observed.

Surface skin-friction lines along with $c_{f,x}$ contours are shown in Fig. 11 for RSM-ln(ω) and RSM-g at the incidence angle of 20°. The predicted flow

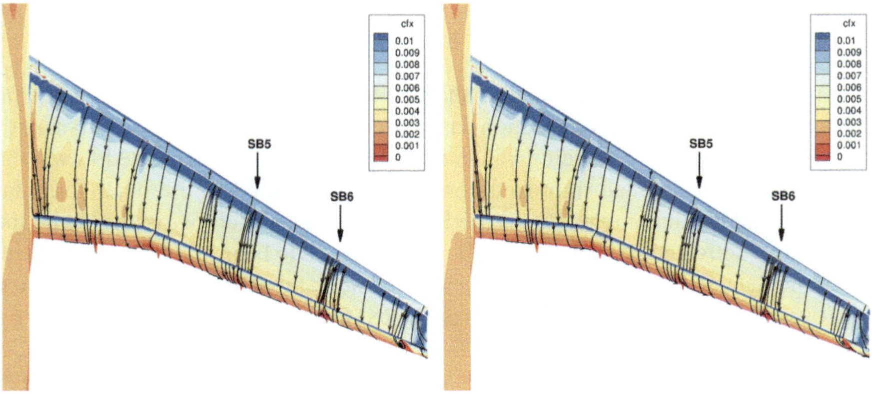

Fig. 11 Surface skin-friction at $\alpha = 20°$

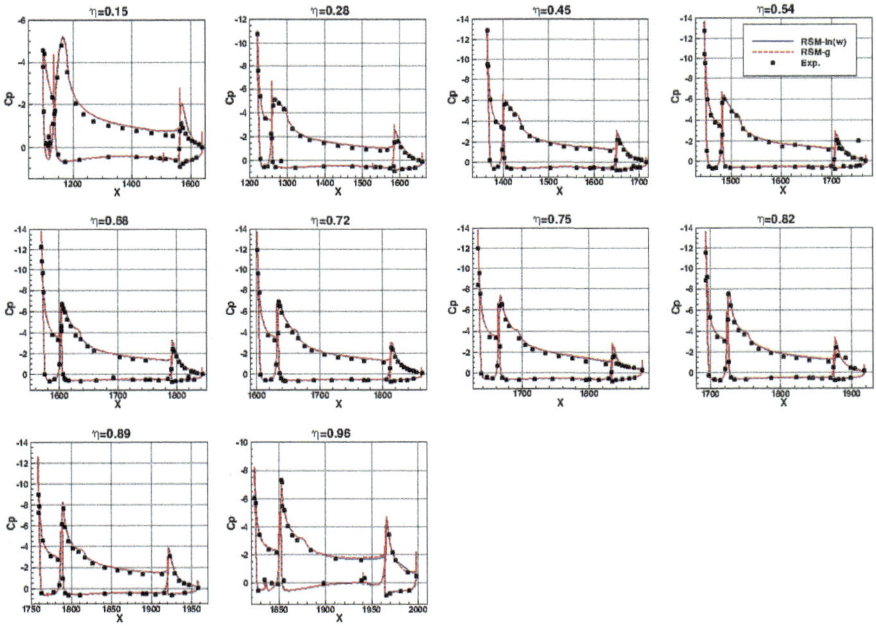

Fig. 12 Sectional c_p at $\alpha = 20°$

topology and c_{fx} levels delivered by both RSM variants are indiscernible. Sectional c_p at different spanwise stations is shown in Fig. 12. At all the stations almost similar pressure distribution by both RSM variants and good agreement with the experimental c_p can be observed.

Overall, for complex high-lift flows RSM-ln(ω) demonstrates better convergence compared to the other RSM variants and shows good agreement with the experimental data.

5 Summary

In this paper, verification, validation and application of RSM-ln(ω) for industrial flows are carried out. For verification case, the RSM-ln(ω) delivered results identical to the RSM-(ω). Concerning validation of RSM-ln(ω) for separated flows, the RSM-ln(ω) demonstrated improved convergence compared to the other formulations. For industrial complex flows, the RSM-(ω) showed stability issues while the RSM-ln(ω) delivered predictions which were observed to be in good agreement with experimental data. For all the cases investigated here, the RSM-ln(ω) demonstrated improved convergence compared to the other RSM variants (RSM-(ω) and RSM-g).

Acknowledgements The funding of partial activities through the Federal Ministry for Economic Affairs and Climate Action under contracts FKZ 20A1504A in the framework of the air transport research program is gratefully acknowledged.

References

1. Spalart, P.R., Allmaras, S.R.: A one-equation turbulence model for aerodynamic flows. In: 30th Aerospace Sciences Meeting & Exhibition (1992)
2. Menter, F.R.: Two-equation Eddy-viscosity turbulence models for engineering applications. AIAA J. **32**, 1598–1605 (1994)
3. Eisfeld, B.: Implementation of Reynolds Stress Models into the DLR-FLOWer Code. Institutsbericht, DLR-IB 124-2004/31. Report of the Institute of Aerodynamics and Flow Technology, Braunschweig, ISSN 1614-7790
4. Eisfeld, B., et al.: Verification and validation of a second-moment-closure model. AIAA J. **54**(5), 1524–1541 (2016)
5. Braun, S.: Implementation of a ln(w)-based SSG/LRR Reynolds stress model into the DLR-TAU Code. Institutsbericht, D LR-IB-AS-BS-2019-37. Report of the Institute of Aerodynamics and Flow Technology, Braunschweig, ISSN 1614-7790
6. Eisfeld, B., et al.: Reynolds-stress model computations of NASA juncture flow experiment. AIAA J. **60**(3), 1643–1662 (2022)
7. Bassi, F., et al.: Discontinuous Galerkin solution of the Reynolds-averaged Navier Stokes and k-ω turbulence model equations. Comput. & Fluids **34** (2005)
8. McCroskey, W.J.: A critical assessment of wind tunnel results for the NACA 0012 airfoil. AGARD CP-429 (1988)
9. NASA Turbulence Modeling Resource website. https://turbmodels.larc.nasa.gov. Accessed June 2018
10. Greenblatt, D., Paschal, K.B., Yao, C.-S., Harris, J.: Experimental investigation of separation control part 2: zero mass-flux oscillatory blowing. AIAA J. **44**(12), 2831–2845 (2006). http://arc.aiaa.org/doi/abs/10.2514/1.19324
11. Togiti, V., Eisfeld, B.: Assessment of g-equation formulation for a second-moment reynolds stress turbulence model. AIAA paper 2015-2925 (2015)
12. Vassberg, J.V., DeHaan, M.A., Rivers, S.M., Wahls, R.A.: Development of a common research model for applied CFD validation studies. In: AIAA-Paper 2008-6919, 26th AIAA Applied Aerodynamics Conference (2008)
13. Keye, S., Togiti, V., Brodersen, O.: DLR Results of the Sixth AIAA Computational Fluid Dynamics Drag Prediction Workshop, 55th AIAA Aerospace Sciences Meeting (2017)

14. Morrison, J.H.: 6th AIAA CFD Drag Prediction Workshop (2016). https://aiaa-dpw.larc.nasa.gov/
15. Rumsey, C.L.: 2nd AIAA CFD High Lift Prediction Workshop (HiLiftPW-2) (2016). https://hiliftpw.larc.nasa.gov
16. Lee-Rausch, E.M., Rumsey, C.L., Eisfeld, B.: Application of a full Reynolds stress model to high lift flows. AIAA paper 2016-3944 (2016)

Aeroelastic AIC-Based Reduced Order Model with CFD-Corrections for Gust Encounter Simulations

Christoph Kaiser and Diliana Friedewald

Abstract An aeroelastic reduced order model for gust encounters of aircraft models is presented. The reduced order model provides aeroelastic time domain responses based on aeroelastic coupling in the frequency domain which is developed into a state space model using rational function approximation. The AIC-based aerodynamics are corrected by quasi-steady RANS solutions in order to improve the aerodynamic response in transonic flow. The correction method is based on a Taylor series expansion of the AIC matrix and allows correcting simultaneously for multiple harmonic pressure responses provided by higher-order methods. The method is presented in detail and discussed for a wing-only configuration. The reduced order model is demonstrated for gust encounters of the NASA Common Research Model in transonic flow and compared to time-linearized RANS-based aeroelastic gust responses.

Keywords Aeroelasticity · Coupling · Gust · Loads · DLM correction · Linearized CFD · LFD · State space · Vector fitting

1 Introduction

The design of aircraft requires a great number of aeroelastic simulations for the evaluation of the relevant loads for aircraft sizing, the assessment of stability and the design of flight control systems. Moreover, the effort of virtual certification increases the number of aeroelastic simulations even more in order to integrate virtual flight testing. Virtual aircraft models, therefore, demand high requirements on both physical accuracy and computational costs.

C. Kaiser (✉) · D. Friedewald
German Aerospace Center, Institute of Aeroelasticity, Göttingen, Germany
e-mail: christoph.kaiser@dlr.de

D. Friedewald
e-mail: diliana.friedewald@dlr.de

R. Heinrich (ed.), *Advanced Aircraft Understanding via the Virtual Aircraft Model*, Notes on Numerical Fluid Mechanics and Multidisciplinary Design 155,
https://doi.org/10.1007/978-3-031-69425-7_4

The doublet lattice method (DLM) [1] provides fast and affordable unsteady aerodynamic responses which are the corner stone for obtaining aeroelastic responses. The DLM is based on potential flow theory and thus, it is limited in accurately describing the unsteady flow physics in important regimes like transonic flow and regimes of flow separation. However, it is possible to numerically enhance the method from aerodynamic responses of higher order closing the gap between the computational costs of the higher-order methods and maintaining the level of physical accuracy. Moreover, the frequency-dependent description of the aerodynamic forces allow for efficient aeroelastic coupling in the frequency domain. The resulting frequency domain model, additionally, can be transformed into the time domain in terms of aeroelastic responses or as a state space model. The latter allows for both efficient time integration and embedding in subsequent time-dependent analysis like control design for load alleviation.

Different approaches for correcting the DLM have been investigated in the past decades. The Transonic DLM method [2] employs an extension of the DLM's kernel by steady flow solutions from CFD. The work of [3] proposes a correction of the inverse of the AIC matrix for a single frequency but accounting for multiple solutions of different mode shapes. In [4], this approach is investigated for building a reduced order model for a frequency bandwidth required by gust encounter simulations. For a comprehensive overview of different correction approaches, the reader is referred to [5].

The proposed correction in this work is based on correction terms introduced in the Taylor series expansion of the governing AIC matrix as presented in [6]. It is extended in this work to account for multiple higher-order solutions resulting from different harmonic motions. Therefore, this paper presents aeroelastic aircraft models for gust encounter simulations which are based on fast DLM but augmented by quasi-steady aerodynamic results from RANS-based CFD. The employed correction method is described in detail and its application is demonstrated for a wing-only configuration as well as for the NASA Common Research Model (CRM) configuration [7].

2 Methods

2.1 Aeroelastic Governing Equations

The aeroelastic coupling is performed in modal coordinates resulting from linear modal analysis of the structural model. This identifies the structural mode shapes and their associated eigenfrequencies. The modal matrix $\mathbf{\Phi}$ comprising the selected mode shapes allows to transform the equations of motion into modal coordinates with the modal mass matrix $\mathbf{M}$ and the modal stiffness matrix $\mathbf{K}$. Furthermore, the equations of motion are transformed into the frequency domain by assuming small harmonic motion and aerodynamic forces with the complex-valued amplitudes $\hat{\mathbf{q}}$ and

$\hat{\mathbf{f}}$, respectively. Taken together, this renders the time-linearized aeroelastic governing equations as

$$\left(-\omega^2\mathbf{M} + \mathbf{K}\right)\hat{\mathbf{q}} = \hat{\mathbf{f}} = \mathbf{Q}(\omega)\hat{\mathbf{q}}\ , \tag{1}$$

where $\mathbf{Q}$ is the generalized aerodynamic force (GAF) matrix. The GAF matrix is the linear transfer function matrix which returns the time-linearized aerodynamic forces $\hat{\mathbf{f}}$ from the amplitudes of the harmonic motion $\hat{\mathbf{q}}$. It is obtained by separately identifying the harmonic aerodynamic responses for the modes shapes of interest and the required frequency bandwidth. By gathering the aerodynamic responses in terms of pressure amplitudes in a frequency-dependent matrix $\hat{\mathbf{C}}_p$, the GAF matrix is obtained as

$$\mathbf{Q}(\omega) = \frac{1}{2}\rho V^2 \mathbf{\Phi}^T \mathbf{H}\hat{\mathbf{C}}_p(\omega)\ ,$$

where $\mathbf{\Phi}^T$ projects the aerodynamic forces onto the modal coordinates and $\mathbf{H}$ is a matrix comprising the aerodynamic force transfer to the structural grid points as well as the areas of the aerodynamic surface discretization. The force transfer is performed by the transpose of the spatial splining matrix, which is required for the transfer of the structural displacements of the mode shapes onto the aerodynamic grid points.

2.1.1 Aeroelastic Response of Gust Encounters

The aerodynamic harmonic response due to modal excitation is obtained from the GAF matrix as described in the previous section. For obtaining the aeroelastic response, the transfer function matrix for the structural displacements has to be coupled with the GAF matrix. Figure 1 shows the aeroelastic feedback loop which couples the aerodynamic forces with the structural model by the structural displacements. This feedback loop describes the aeroelastic equations of Eq. (1) but extended by an additional aerodynamic excitation $\hat{\mathbf{q}}_{gust}(\omega)$ representing gust encounters. Therefore, the GAF matrix must be expanded by an extra column containing the aerodynamic responses due to sinusoidal gusts. The GAF matrix, thus, introduces one additional degree of freedom, which the structural model has to numerically handle as zero in the following equations. Hence, gust encounters are incorporated in a time-linearized way by their frequency spectrum.

The feedback loop of Fig. 1 allows to derive the transfer function matrix depending on the specified input and output. For gust encounter simulations, the input is the gust excitation $\hat{\mathbf{q}}_{gust}$ and for the output it is either the modal amplitudes $\hat{\mathbf{q}}$ or the forces $\hat{\mathbf{f}}$. The first case is obtained to

$$\hat{\mathbf{q}}(\omega) = \left(\mathbf{K} - \omega^2\mathbf{M} - \mathbf{Q}(\omega)\right)^{-1}\mathbf{Q}(\omega)\hat{\mathbf{q}}_{gust}(\omega) \tag{2}$$

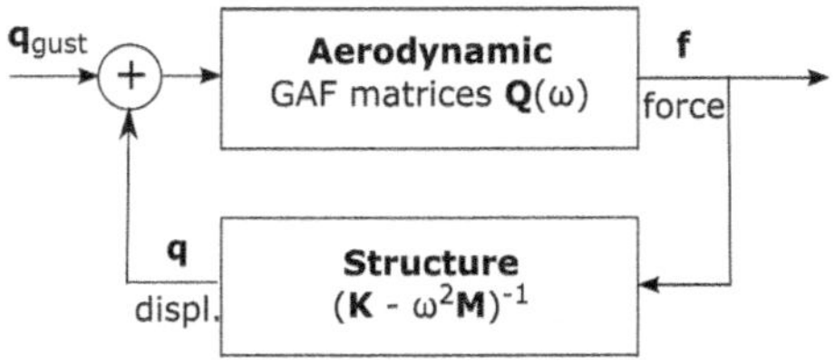

Fig. 1 Aeroelastic feedback loop for gust encounters in the frequency domain

from Eq. (1) by solving for $\hat{\mathbf{q}}$ with the additional excitation on the right-hand side. The latter case requires to apply Eq. (1), which is $\hat{\mathbf{f}} = (\mathbf{K} - \omega^2\mathbf{M})\hat{\mathbf{q}}$, once again to get

$$\hat{\mathbf{f}}(\omega) = \left(\mathbf{I} - \mathbf{Q}(\omega)\left(\mathbf{K} - \omega^2\mathbf{M}\right)^{-1}\right)^{-1}\mathbf{Q}(\omega)\hat{\mathbf{q}}_{gust}(\omega)\,. \tag{3}$$

With these two transfer function matrices the aeroelastic responses for a given set of frequencies in modal coordinates are obtained. If the aerodynamic transfer-function matrices are available for the surface force distribution, these equations also allow for load section recovery as performed in [8]. Furthermore, the aeroelastic frequency responses can be transformed into the time domain by the inverse Fourier transform or a state space model can be fitted as described in the following section.

2.1.2 Aeroelastic State Space Model

Time-domain analysis like control design requires the previously defined aeroelastic transfer function matrices to be transformed from the frequency domain into the time domain. For this, the approach of rational function approximation (RFA) allows finding a rational function which approximately fits a sampled function in the frequency domain. The poles and zeros of the RFA, then, provide a direct way to identify the system matrices **A**, **B**, **C** and **D** of a state space model. Therefore, the the aeroelastic transfer function matrix is sampled in the frequency domain and the process of RFA is applied. In this work, the Vector Fitting Method [9] is used in order to determine the poles and zeros for each sampled column of the transfer function matrix. The process requires to specify the order of the rational function which determines the (equal) number of poles and zeros. The Vector Fitting Method is able to return only stable poles which is important in the subsequent time integration of the state space model. The order is obtained by finding a low order representation of the sampled function which minimizes the error between the sampled function and the rational function to a suitable degree.

The resulting state space model is of the form

$$\begin{aligned}\dot{\mathbf{x}} &= \mathbf{A}\mathbf{x} + \mathbf{B}\mathbf{u}\\ \mathbf{y} &= \mathbf{C}\mathbf{x} + \mathbf{D}\mathbf{u}\,,\end{aligned}$$

where $\mathbf{x}$ are the states and $\mathbf{u}$ and $\mathbf{y}$ are the input and output vectors as specified by the transfer function matrix of the frequency domain. The states do not directly correspond to physical degrees of freedom, since they are a result of the fitting process. The obtained state space model is checked by comparing its time domain response with the frequency response of the original transfer function matrix by transforming the frequency response into the time domain. The time integration of the state space model is performed by a fourth order Runge–Kutta scheme.

2.2 Aerodynamic Governing Equations

The governing equation of DLM relates local pressure differences to locally induced downwash. By assuming harmonic motion, the harmonic amplitudes are related by a linear transformation with the matrix of aerodynamic influence coefficients (AIC). The downwash amplitudes $\hat{\mathbf{w}}$ results from the substantial derivative of a given harmonic motion. Therefore, the amplitudes of the pressure differences $\Delta\hat{\mathbf{c}}_p$ are obtained by solving

$$\hat{\mathbf{w}} = \mathbf{AIC}(\mathrm{Ma}, \omega^*)\Delta\hat{\mathbf{c}}_p \ . \tag{4}$$

The aerodynamic influence matrix **AIC**, thus, describes the fluidic influence between the oscillating parts of the wing discretization based on potential flow theory. It defines the sending and receiving characteristics of the interrelation between the DLM boxes. Each box induces a downwash depending on its corresponding pressure difference, which is in turn affected by the induced downwash from the other boxes. This is due to the fact that the boundary conditions at each box do not allow flow through. The aerodynamic influence matrix especially depends on the Mach number Ma and the reduced frequency ω^* of the harmonic motion, where the reduced frequency is defined as

$$\omega^* = \omega \frac{L}{V}$$

with L being the characteristic length of the configuration and V the freestream velocity.

In the following, the Mach number is dropped from the notation as it is considered as constant.

2.2.1 Correction of Aerodynamic Influence Matrix

The objective of a DLM correction is to improve the aerodynamic influence matrix **AIC** from aerodynamic responses obtained from higher-order methods. In order to implement a correction of the **AIC** matrix, a correction of the Taylor series of the **AIC** matrix is proposed. The Taylor series around $\omega^* = 0$ is written as

$$\mathbf{AIC}(\omega^*) = \mathbf{AIC}_0\big|_{\omega^*=0} + \mathbf{AIC}_1\big|_{\omega^*=0}\omega^* + \mathbf{AIC}_2\big|_{\omega^*=0}\omega^{*2} + O\left(\omega^{*3}\right) , \tag{5}$$

where $\mathbf{AIC}_k$ are the partial derivatives of order k of the **AIC** matrix with respect to ω^*:

$$\mathbf{AIC}_k\big|_{\omega^*} = \frac{1}{k!}\frac{\partial^k \mathbf{AIC}(\omega^*)}{\partial \omega^{*k}} .$$

The **AIC** matrix is then corrected by introducing additive matrices for each term of the Taylor series, which yields the corrected matrix

$$\begin{aligned}\widetilde{\mathbf{AIC}}(\omega^*) &= (\mathbf{AIC}_0\big|_{\omega^*=0} + \mathbf{C}_0) + (\mathbf{AIC}_1\big|_{\omega^*=0} + \mathbf{C}_1)\omega^* + (\mathbf{AIC}_2\big|_{\omega^*=0} + \mathbf{C}_2)\omega^{*2} \\ &\quad + O\left(\omega^{*3}\right) \\ &= \mathbf{AIC}(\omega^*) + \mathbf{C}_0 + \mathbf{C}_1\omega^* + \mathbf{C}_2\omega^{*2} .\end{aligned}$$

The latter equation is found from identifying the terms of Eq. (5) and only considering three correction matrices. Especially, if only the correction matrix for the zeroth order term is considered, the quasi-steady corrected matrix

$$\widetilde{\mathbf{AIC}}(\omega^*) = \mathbf{AIC}(\omega^*) + \mathbf{C}_0 \tag{6}$$

is obtained. Although the correction matrix $\mathbf{C}_0$ is applied linearly for all frequencies, it is noted that the effect on the pressure difference amplitudes is obtained by the inverse of $\widetilde{\mathbf{AIC}}(\omega^*)$. In general, a correction of the Taylor series of the **AIC** matrix around $\omega^* = 0$ is only valid in a region close to $\omega^* = 0$. This requirement is approximately fulfilled for applications in aeroelasticity investigated here, where the structural frequencies of interest are small. In this work only the quasi-steady correction is investigated.

Inserting the quasi-steady corrected matrix in the DLM equation (4), and evaluating at $\omega^* = 0$ yields

$$\hat{\mathbf{w}}\big|_{\omega^*=0} = (\mathbf{AIC}\big|_{\omega^*=0} + \mathbf{C}_0)\Delta\hat{\mathbf{c}}_p\big|_{\omega^*=0} . \tag{7}$$

This equation allows to determine the correction matrix $\mathbf{C}_0$, if the pressure difference amplitudes from a higher-order method, $\Delta\hat{\mathbf{c}}_p$, are provided for one given quasi-steady motion. The term motion comprises both harmonic motion from mode shapes as well as sinusoidal gusts. Equation (7) is an underdetermined system of n equations for the n^2 unknown entries of $\mathbf{C}_0$.

A special case of the proposed correction is a diagonal scaling matrix $\mathbf{D}$ for the **AIC** matrix as it is investigated in [6, 10, 11]. The restriction to a diagonal matrix allows to obtain this matrix directly from the corrected DLM equation at $\omega^* = 0$,

$$\hat{\mathbf{w}}\big|_{\omega^*=0} = \mathbf{D}\,\mathbf{AIC}\big|_{\omega^*=0}\Delta\hat{\mathbf{c}}_p\big|_{\omega^*=0} ,$$

by solving for each entry of the diagonal of $\mathbf{D}$ separately. In particular, there are n unknowns $\mathbf{D}_{ii}$ and n equation from the pressure amplitudes of one quasi-steady motion. This scaling matrix affects the receiving property of each DLM box by scaling the rows of the $\mathbf{AIC}$ matrix. The multiplicative matrix $\mathbf{D}$ can be expressed as the additive correction matrix

$$\mathbf{C}_0 = (\mathbf{D} - \mathbf{I})\mathbf{AIC}\big|_{\omega^*=0}$$

of Eqs. (6) and (7).

A more general case is achieved by correcting the $\mathbf{AIC}$ matrix considering results from multiple quasi-steady motions. This is especially important for aeroelastic applications with multiple modal degrees of freedom and it can further comprise harmonic gusts. For this, Eq. (7) is extended for these multiple quasi-steady motions. Due to the linearity of the underlying Eq. (4), the vectors of amplitudes are substituted by matrices of m column vectors of amplitudes for m quasi-steady motions yielding

$$\hat{\mathbf{W}}\big|_{\omega^*=0} = (\mathbf{AIC}\big|_{\omega^*=0} + \mathbf{C}_0)\Delta\hat{\mathbf{C}}_p\big|_{\omega^*=0} \,. \tag{8}$$

This corresponds to nm equations for the unknown entries of the correction matrix $\mathbf{C}_0$. In general, it is $m \leq n$ resulting in an underdetermined problem. This equation can be solved for $\mathbf{C}_0$ by the help of the Moore–Penrose inverse of the matrix of pressure difference amplitudes, denoted by $^+$. The proposed correction matrix is

$$\mathbf{C}_0 = (\hat{\mathbf{W}}\big|_{\omega^*=0} - \mathbf{AIC}\big|_{\omega^*=0}\Delta\hat{\mathbf{C}}_p\big|_{\omega^*=0})\Delta\hat{\mathbf{C}}_p^+\big|_{\omega^*=0} \,, \tag{9}$$

which fulfills Eq. (8), if $\Delta\hat{\mathbf{C}}_p^+\Delta\hat{\mathbf{C}}_p = \mathbf{I}$ and thus, if $\Delta\hat{\mathbf{C}}_p$ has full column rank:

$$\begin{aligned}
\hat{\mathbf{W}}\big|_{\omega^*=0} &= \left[\mathbf{AIC}\big|_{\omega^*=0} + \left(\hat{\mathbf{W}}\big|_{\omega^*=0} - \mathbf{AIC}\big|_{\omega^*=0}\Delta\hat{\mathbf{C}}_p\big|_{\omega^*=0}\right)\Delta\hat{\mathbf{C}}_p^+\big|_{\omega^*=0}\right]\Delta\hat{\mathbf{C}}_p\big|_{\omega^*=0} \\
&= \mathbf{AIC}\big|_{\omega^*=0}\Delta\hat{\mathbf{C}}_p\big|_{\omega^*=0} + \hat{\mathbf{W}}\big|_{\omega^*=0} - \mathbf{AIC}\big|_{\omega^*=0}\Delta\hat{\mathbf{C}}_p\big|_{\omega^*=0} \\
&= \hat{\mathbf{W}}\big|_{\omega^*=0} \,.
\end{aligned}$$

Equation (9) is also obtained by the vectorization of Eq. (8), which results in an underdetermined linear system of equations for the vectorized correction matrix $\mathbf{C}_0$. First, Eq. (8) is written as

$$\hat{\mathbf{W}}\big|_{\omega^*=0} - \mathbf{AIC}\big|_{\omega^*=0}\Delta\hat{\mathbf{C}}_p\big|_{\omega^*=0} = \mathbf{I}\mathbf{C}_0\Delta\hat{\mathbf{C}}_p\big|_{\omega^*=0}$$

and then, it is vectorized with the help of the Kronecker product, denoted $\otimes$, and its application to the matrix multiplication[1] on the right-hand side:

[1] $\mathrm{vec}\,(ABC) = (C^T \otimes A)\mathrm{vec}\,(B)$.

$$\mathrm{vec}\left(\hat{\mathbf{W}}\big|_{\omega^*=0} - \mathbf{AIC}\big|_{\omega^*=0}\Delta\hat{\mathbf{C}}_p\big|_{\omega^*=0}\right) = \left(\Delta\hat{\mathbf{C}}_p^T\big|_{\omega^*=0} \otimes \mathbf{I}\right) \mathrm{vec}\left(\mathbf{C}_0\right) .$$

Subsequently, the ordinary least squares solution vec $(\mathbf{C}_0)$ is found by the Moore–Penrose inverse of the Kronecker product:

$$\begin{aligned}
\mathrm{vec}\left(\mathbf{C}_0\right) &= \left(\Delta\hat{\mathbf{C}}_p^T\big|_{\omega^*=0} \otimes \mathbf{I}\right)^+ \mathrm{vec}\left(\hat{\mathbf{W}}\big|_{\omega^*=0} - \mathbf{AIC}\big|_{\omega^*=0}\Delta\hat{\mathbf{C}}_p\big|_{\omega^*=0}\right) \\
&= \left((\Delta\hat{\mathbf{C}}_p^+\big|_{\omega^*=0})^T \otimes \mathbf{I}\right) \mathrm{vec}\left(\hat{\mathbf{W}}\big|_{\omega^*=0} - \mathbf{AIC}\big|_{\omega^*=0}\Delta\hat{\mathbf{C}}_p\big|_{\omega^*=0}\right) \\
&= \mathrm{vec}\left(\mathbf{I}\left(\hat{\mathbf{W}}\big|_{\omega^*=0} - \mathbf{AIC}\big|_{\omega^*=0}\Delta\hat{\mathbf{C}}_p\big|_{\omega^*=0}\right)\Delta\hat{\mathbf{C}}_p^+\big|_{\omega^*=0}\right) .
\end{aligned}$$

The latter equation equals Eq. (9). It is obtained by writing the right-hand side again as a matrix multiplication after taking the Moore–Penrose inverse of the Kronecker factors and interchanging transposition and inverting. The found least squares solution is not unique, since the Kronecker product of above equation has less rows than columns ($nm \leq n^2$) and thus, it does not have full column rank. Consequently, Eq. (9) for the quasi-steady correction matrix provides a least squares solution for Eq. (8), which minimizes the Frobenius norm of $\mathbf{C}_0$. However, in the expected case where $\Delta\hat{\mathbf{C}}_p$ has full column rank it fulfills Eq. (8).

Finally, the process of determining the correction matrices of higher order is outlined. The higher-order correction matrices $\mathbf{C}_i$ are obtained by successively correcting the **AIC** matrix. As in the quasi-steady case, the starting point is Eq. (4) with the corrected matrix $\widetilde{\mathbf{AIC}}$ applied, cf. Eq. (8):

$$\hat{\mathbf{W}}\big|_{\omega^*} = (\mathbf{AIC}\big|_{\omega^*} + \mathbf{C}_0 + \mathbf{C}_1\omega^*)\Delta\hat{\mathbf{C}}_p\big|_{\omega^*} .$$

The differentiation of this equation with respect to ω^* gives

$$\frac{\partial\hat{\mathbf{W}}}{\partial\omega^*}\bigg|_{\omega^*} = \left(\frac{\partial\mathbf{AIC}}{\partial\omega^*}\bigg|_{\omega^*} + \mathbf{C}_1\right)\Delta\hat{\mathbf{C}}_p\big|_{\omega^*} + \left(\mathbf{AIC}\big|_{\omega^*} + \mathbf{C}_0 + \mathbf{C}_1\omega^*\right)\frac{\partial\Delta\hat{\mathbf{C}}_p}{\partial\omega^*}\bigg|_{\omega^*}$$

and the evaluation at $\omega^* = 0$ yields the equation for $\mathbf{C}_1$:

$$\frac{\partial\hat{\mathbf{W}}}{\partial\omega^*}\bigg|_{\omega^*=0} = \left(\frac{\partial\mathbf{AIC}}{\partial\omega^*}\bigg|_{\omega^*=0} + \mathbf{C}_1\right)\Delta\hat{\mathbf{C}}_p\big|_{\omega^*=0} + (\mathbf{AIC}\big|_{\omega^*=0} + \mathbf{C}_0)\frac{\partial\Delta\hat{\mathbf{C}}_p}{\partial\omega^*}\bigg|_{\omega^*=0}$$

Then, setting the first-order correction matrix to

$$\mathbf{C}_1 = \left(\frac{\partial\hat{\mathbf{W}}}{\partial\omega^*}\bigg|_{\omega^*=0} - \frac{\partial\mathbf{AIC}}{\partial\omega^*}\bigg|_{\omega^*=0}\Delta\hat{\mathbf{C}}_p\big|_{\omega^*=0} - (\mathbf{AIC}\big|_{\omega^*=0} + \mathbf{C}_0)\frac{\partial\Delta\hat{\mathbf{C}}_p}{\partial\omega^*}\bigg|_{\omega^*=0}\right)\Delta\hat{\mathbf{C}}_p^+\big|_{\omega^*=0}$$

fulfills above equation if $\Delta\hat{\mathbf{C}}_p$ has full column rank and it is $\Delta\hat{\mathbf{C}}_p^+\Delta\hat{\mathbf{C}}_p = \mathbf{I}$. All derivatives are obtained by central finite differences at $\omega^* = 0$ exploiting the symmetry of the complex-valued functions for negative frequencies. However, this requires an additional solution from the higher-order method very close to the frequency of zero in order to perform the finite difference for $\partial \Delta \hat{\mathbf{C}}_p / \partial \omega^* \big|_{\omega^*=0}$.

For determining $\mathbf{C}_2$, the above process is repeated but the corrected DLM equation is differentiated twice. Consequently, derivatives of second order are required. Continuing this process allows to determine many correction matrices $\mathbf{C}_i$, however, for the cost of providing additional solutions from higher-order methods. Moreover, higher-order correction matrices for the Taylor series significantly distort the results if evaluated for frequencies considerably greater than zero. Taken together, only corrections up to the first order appear useful for building an efficient reduced order model.

2.2.2 Geometric Mapping of CFD Surface Mesh

The proposed AIC correction method requires solutions of harmonic pressure difference amplitudes from higher-order methods. In general, higher-order methods allow a more detailed discretization of the configuration than the DLM method. Moreover, higher-order methods like CFD methods provide surface pressure distributions which must be converted to pressure differences to be used in the AIC correction. Hence, a prerequisite for the correction is a process which obtains pressure differences for each DLM box from the higher-order solutions accounting for geometrical differences and available quantities. Due to these differences, only approximations of the higher-order solutions can be obtained by mapping these solutions on the DLM grid. Thus, different approaches for such a mapping process are possible depending on selected constraints like local accuracy or accuracy of global forces. Figure 2 displays the CFD surface mesh and the DLM grid for the NASA CRM configuration illustrating the geometrical differences between both methods.

In this work, the mapping process returns the pressure differences for each DLM box by approximating the surface pressure distribution accordingly to the DLM discretization and by separating the surface cells into upper and lower surface cells. The latter allows computing the pressure differences by subtracting the upper from the

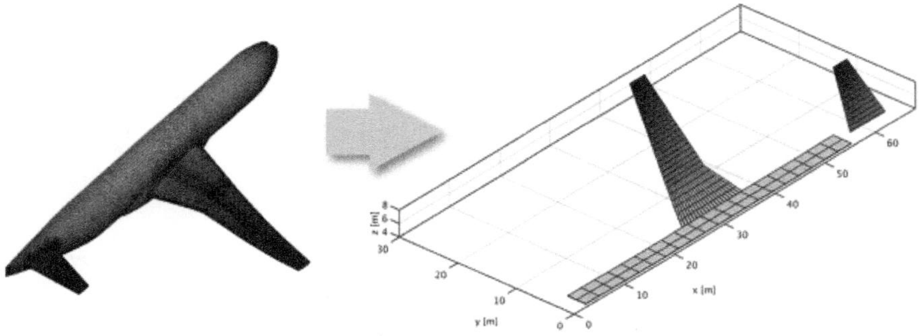

Fig. 2 CFD surface mesh and DLM grid of the NASA CRM half model

lower surface pressures. The terms upper and lower surface result from the definition that the DLM box's normal vector points to the upper surface. Therefore, the association of the surface cells to the upper and lower surface is performed by comparing the normal vectors of the surface cells to the normal vectors of the attributed DLM box. If the surface cell normals are defined to point outside the fluid domain, then the cells associated with the upper surface have normal vectors pointing in the inverse direction of the DLM box's normal vector. Besides this automatic association, a manual association may also be feasible. For example, if the boundary markers are already denoted as the upper or lower surface in the process of mesh generation.

For obtaining the pressure values for each DLM box, the surface pressure distribution is interpolated employing Gaussian Quadrature. Gaussian Quadrature of quadrilaterals defines integration points as well as corresponding weights for each DLM box depending on the integration order. The integration points are used to sample the surface pressure distribution. By dividing the integration weights with the area of the standard interval, the integration is rendered as a weighted averaging of the surface pressure distribution over a DLM box. The integration order is increased to reflect the number of surface nodes which can be attributed to a DLM box. For this, the surface nodes are projected onto the plane spanned by a DLM box and the number of nodes falling inside the box are counted. By contrast, the integration points are projected along the DLM box's normal direction onto the surface cells associated with both the upper and lower surface. Hence, each integration point is assigned two pressure values to define the pressure difference. They are obtained from the surface cells where the projected integration points fall inside the cells. The pressure values, then, result from the weighted sum of the nodal values using barycentric coordinates.

This process of mapping the surface pressures of the CFD solution onto the DLM grid is illustrated in Fig. 3. Furthermore, the process is expressed as a sparse mapping matrix $\mathbf{J}$. This matrix comprises the factors of Gaussian weights, barycentric weights and a sign reflecting the upper or lower surface for each surface node and DLM box. Thus, it renders the pressure difference amplitudes from the amplitudes of the pressure distribution as

$$\Delta\hat{\mathbf{C}}_p(\omega^*) = \mathbf{J}\hat{\mathbf{C}}_p(\omega^*) \; .$$

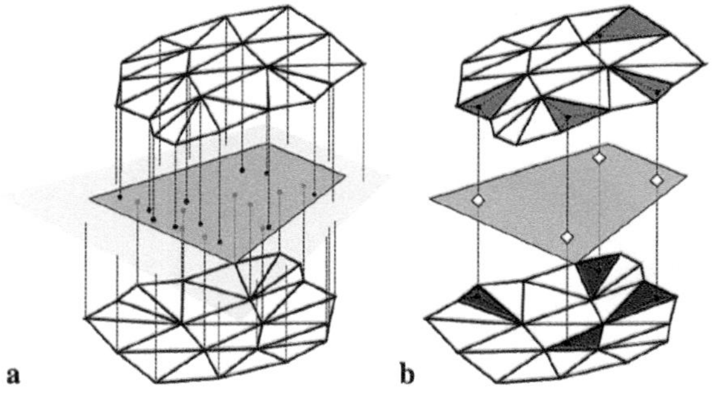

Fig. 3 Illustration of the mapping process: **a** counting projected CFD surface nodes inside a DLM box, **b** projection of the integration points onto CFD surface cells

3 Results

3.1 Aerostabil Test Case

In this test case, the correction method is demonstrated for the aerodynamic response of the Aerostabil configuration. It is a wing-only configuration developed for the wind-tunnel test campaign of the Aerostabil experiment [12]. The swept wing is tapered with an aspect ratio of 3.3 and it has a mean aerodynamic chord of 0.183 m. The numerical model consists of a clamped FEM shell model with 2471 elements, a DLM grid with 520 boxes and a CFD grid with $2.4 \cdot 10^6$ nodes inside a farfield hemisphere of 24 m.

Figure 4 shows the DLM grid as well as the first three mode shapes by means of the deformed DLM grid. The first mode shape is the first bending mode, while the second mode shape is a torsional and bending mode. And the third mode shape is the second bending mode. All mode shapes also exhibit small local deformations close to the wing root. The corresponding eigenfrequencies of the mode shapes as well as the flow parameters for the test case are listed in Table 1.

The investigated test case is in the transonic regime at a Mach number of 0.8 and zero angle of attack. The steady flow field is computed with the DLR TAU-Code [13] using the Spalart-Allmaras turbulence model [15]. The DLR TAU-Code is an unstructured finite-volume solver for the RANS equations equipped with the ALE extension for moving-grid simulations [14]. Based on the steady solution, the Linear Frequency Domain (LFD) solver of the DLR TAU-Code provides the time-linearized unsteady RANS solution to a given harmonic motion [16, 17]. Moreover, the LFD solver returns the amplitudes of the surface pressure distribution which are used for the proposed correction of the AIC matrix. Hence, the quasi-steady correction matrix $\mathbf{C}_0$ is obtained for LFD solutions at a frequency of zero following Sects. 2.2.1 and 2.2.2. The AIC matrix is computed by an in-house DLM tool. Then, the corrected

Fig. 4 First three mode shapes of the Aerostabil model splined onto the DLM grid

Table 1 Parameter settings for the Aerostabil test case

Parameter	Symbol	Value
Mach number	Ma	0.8
Reynolds number	Re_L	$1.8 \cdot 10^6$
Angle of attack	α	0°
Free-stream velocity	V	160.35 m/s
Characteristic length	L	0.183 m
1st eigenfrequency	f_1	40.19 Hz
2nd eigenfrequency	f_2	51.27 Hz
3rd eigenfrequency	f_3	116.67 Hz

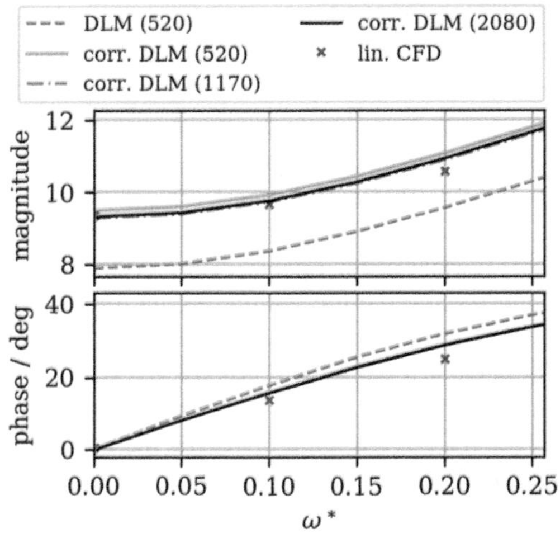

Fig. 5 Derivative of the lift coefficient regarding to the first mode shape

AIC matrix is employed to compute the unsteady aerodynamic responses in terms of the lift coefficient derivatives. Besides the quasi-steady solution, the LFD solver is employed at higher frequencies for obtaining the reference solutions.

In a first step, the corrected DLM response for the first mode shape is compared to the DLM solution as well as the LFD solutions at selected reduced frequencies. The quasi-steady correction matrix is obtained from the LFD solution for the first mode shape at the frequency of zero. Figure 5 shows the influence of the DLM grid discretization on the lift response together with the DLM and LFD responses. By definition, the mapped result at the frequency of zero is reproduced by the corrected AIC matrix. Consequently, the quasi-steady lift derivative of the LFD solution is well rendered by the corrected DLM solution. With increasing numbers of DLM boxes, the difference becomes smaller since the surface distribution is better approximated. In comparison to the DLM solution, the quasi-steady corrected DLM solution matches the LFD solutions much better in the displayed region. Moreover, the phase of the

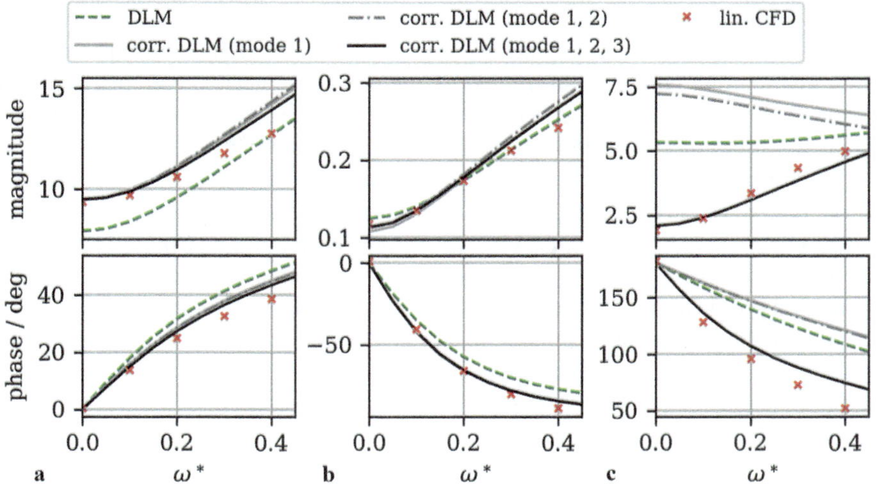

Fig. 6 Derivatives of the lift coefficient regarding to the **a** first mode shape, **b** second mode shape, and **c** third mode shape

lift response is also slightly improved which is due to the effect of the inverse of the quasi-steady corrected AIC matrix on the frequency response as described in Sect. 2.2.1.

The effect of the mode shapes considered in the DLM correction is investigated in Fig. 6. It shows the aerodynamic responses to the first three modes for the DLM and LFD solutions as well as for three corrected DLM solutions using the grid with 520 DLM boxes. The three corrected DLM solutions differ by the number of mode shapes which are considered in the quasi-steady correction matrix. Likewise to Fig. 5, the first correction uses only the response to the first mode shape in the correction method. The second correction also incorporates the second mode shape and the quasi-steady solutions for all three mode shapes are included in the third correction.

The greatest differences between the employed methods are observed in the aerodynamic responses to the third mode, see Fig. 6c. While the third correction renders the reference solution well, in both magnitude and phase, the first and second corrections alter the DLM solution in the opposite direction. This shows that the corrections including the first two mode shapes do not improve the response to the third mode, but the response is affected by the corrections. For the response to second mode, Fig. 6b, all three corrections provide similar responses. Hence, the first correction improves the response to the second mode although it is not included in the correction. This is explainable by the similarity of the first two mode shapes as previously described. In contrast, the third mode shape differs strongly from the first two mode shapes. Thus, the first and second corrections do not contain the information necessary for the correction of the third mode shape which leads to the deterioted results of the first and second correction in Fig. 6c.

Furthermore, the response to the second mode of the third correction provides almost the same results as for the first and second corrections. Subsequently, the additional correction of the third mode shape does not interfere significantly with the other corrections. This is also observed in the response to the first mode, Fig. 6a, where all three corrections provide very similar responses.

Taken together, the results for the Aerostabil wing show that for obtaining the aeroelastic response employing the proposed correction, all mode shapes considered have to be included in the process of creating the correction matrix. For very similar mode shapes, only one mode shape can be sufficient. Moreover, the proposed correction proofs to be suitable for considering multiple mode shapes at once, since the incorporation of additional mode shapes do not significantly alter the response to the other modes. Additionally, all corrections exactly return the mapped responses at the frequency of zero as it is required by the correction process.

3.2 CRM Test Case

In this test case, the proposed aeroelastic reduced order model is investigated for the CRM configuration with gust encounters. The aeroelastic responses due to a 1-cos gust encounter are compared for the DLM and time-linearized RANS solutions as well as for the proposed DLM correction and the fitted state space model including the DLM correction.

The 1-cos gust encounter is applied at an altitude of 9133 m in cruise condition of the CRM configuration which is in transonic flow at $Ma = 0.86$ and $Re_L = 56.27 \cdot 10^6$. For this flight point, the CFD mesh [18] is available in flight shape for longitudinal flight at an angle of attack of $1.641°$. The DLR TAU-Code with its LFD solver is employed for obtaining the steady and time-linearized unsteady RANS solutions likewise to the test case of the Aerostabil wing. The CFD mesh is a half-model consisting of $3.7 \cdot 10^6$ nodes in a hemisphere of around 750 m. Figure 2 depicts both the CFD surface mesh and the DLM grid consisting of 800 boxes. The great differences in the modelled details of the CRM geometry illustrates the approximative properties of the mapping process. This applies especially to the simplified modelling of the fuselage and connection to the horizontal tail plane which is justified only for the investigated longitudinal flight. The condensed dynamic model FERMAT-C2 [19] for the CRM configuration provides the structural properties and the mode shapes for the aeroelastic model. In this test case, two rigid body motions, heave and pitch, as well as the first mode shape are considered. The mode shape is the first symmetric wing bending. Table 2 summarizes both the structural and the flow parameters for this test case. The vertical 1-cos gust encounter has a length of 210 m and an amplitude of 0.625 m/s. The very small gust amplitude emphasizes the time-linearized properties of the employed aerodynamic methods.

The corrected DLM is obtained from the quasi-steady LFD solutions for the pitching motion and the first mode shape. The heave motion does not provide a quasi-steady contribution and the response to the quasi-steady harmonic gust motion

Table 2 Parameter settings for the CRM test case

Parameter	Symbol	Value
Mach number	Ma	0.86
Reynolds number	Re_L	$56.27 \cdot 10^6$
Angle of attack	α	1.641°
Free-stream velocity	V	260.7 m/s
Characteristic length	L	7 m
Mass	m	$2.6 \cdot 10^5$ kg
Moment of inertia	I_y	$2.58 \cdot 10^7$ kg m^2
1st eigenfrequency	f_1	1.057 Hz

is equal to the quasi-steady pitching response. For the corrected DLM solutions, the aeroelastic transfer functions are finely sampled for a low frequency bandwidth up to a reduced frequency of 0.35. These are used to fit a state space model to each of the two transfer functions, for the generalized forces and displacements, employing the process of Sect. 2.1.2 with a chosen order of 30. Time integration with a fourth order Runge–Kutta scheme of the state space models returns the aeroelastic responses in force and displacement due to the 1-cos signal which is applied as the gust input. The obtained responses of the state space models are compared to the time domain responses obtained from the aeroelastic frequency responses by the inverse Fourier transform. The frequency responses result from the convolution of the transfer functions with the spectrum of the 1-cos gust signal for the three considered aerodynamic methods DLM, LFD and corrected DLM.

Figures 7 and 8 show the aeroelastic responses in terms of the generalized forces and displacements due to the 1-cos gust encounter. For the heave and bending force the corrected DLM is in very good agreement with the response using the time-linearized RANS solutions. The proposed correction improves the DLM-based response in the important first peak as well as in the following amplitudes. For the pitching moment the correction returns attenuated first and second peaks while the following amplitudes matches well the time-linearized RANS-based response. The state space model renders the almost identical response for all three forces as the transformed frequency response for the corrected DLM. The generalized displacements in Fig. 8 show a significant improvement of the DLM-based response employing the proposed correction. Likewise to the generalized forces, the heave and bending response are well captured by the corrected DLM method while for the pitching motion the greatest differences are observed. In contrast to the pitching moment, the pitching motion is closer to the time-linearized RANS-based response and the improvement of the DLM-based response is observed. The response of the state space model is again almost equal to the transformed frequency response although a very small offset in the heave displacements for later times is noticeable.

The investigated aeroelastic responses of the CRM configuration due to the 1-cos gust encounter reveals that the interaction of the aerodynamic transfer functions

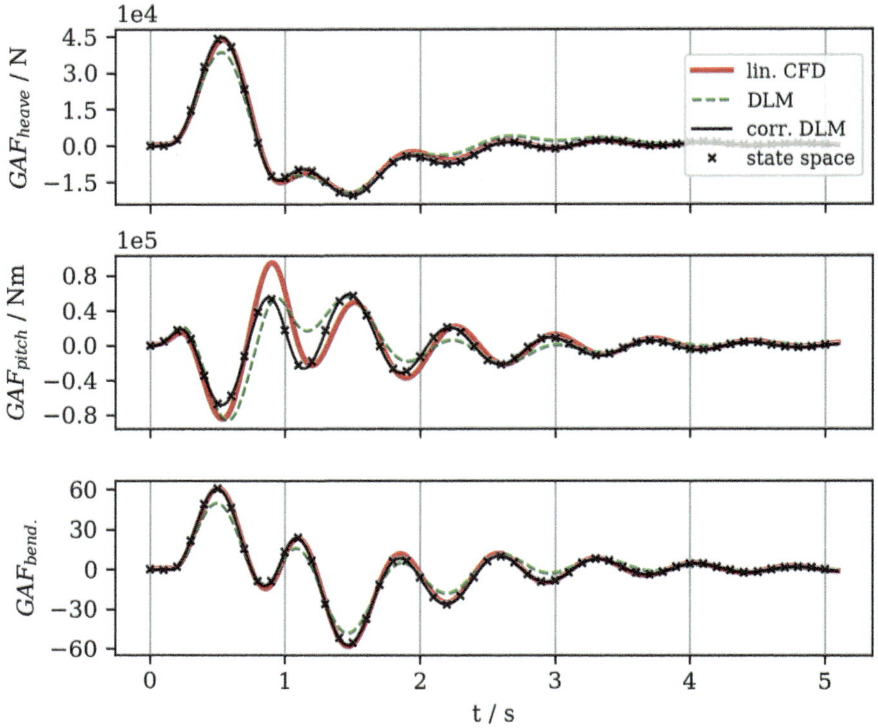

Fig. 7 Aeroelastic response of the CRM due to 1-cos gust encounter in terms of generalized forces for a gust length of 210 m

with the structural dynamic model has a significant effect on the final responses. The corrected DLM uses higher-order aerodynamic responses for the quasi-steady pitching and bending motion. However, the aeroelastic responses show improvements in the heave and bending coordinates and less improvement in the pitch coordinate. The difference in the improvement of the generalized forces and displacements must be related to the additional convolution of the structural transfer function in the case of the generalized forces. However, further investigations have to be carried out. This includes the poorer improvement of the responses in pitch coordinate.

For smaller gust lengths, similar results are obtained as it is exemplarily shown in Fig. 9 for a gust length of 10 m. The same state space model is employed as well as the same aeroelastic transfer functions. In comparison to the longer gust, the shorter gust results in a quicker but attenuated response. Again, the corrected DLM shows very good improvements in the heave and bending motion while the second peak of the pitching motion is underpredicted. For shorter gusts, the frequency bandwidth of the excitation is increased to higher frequencies which needs to be considered in the fitted bandwidth for the RFA.

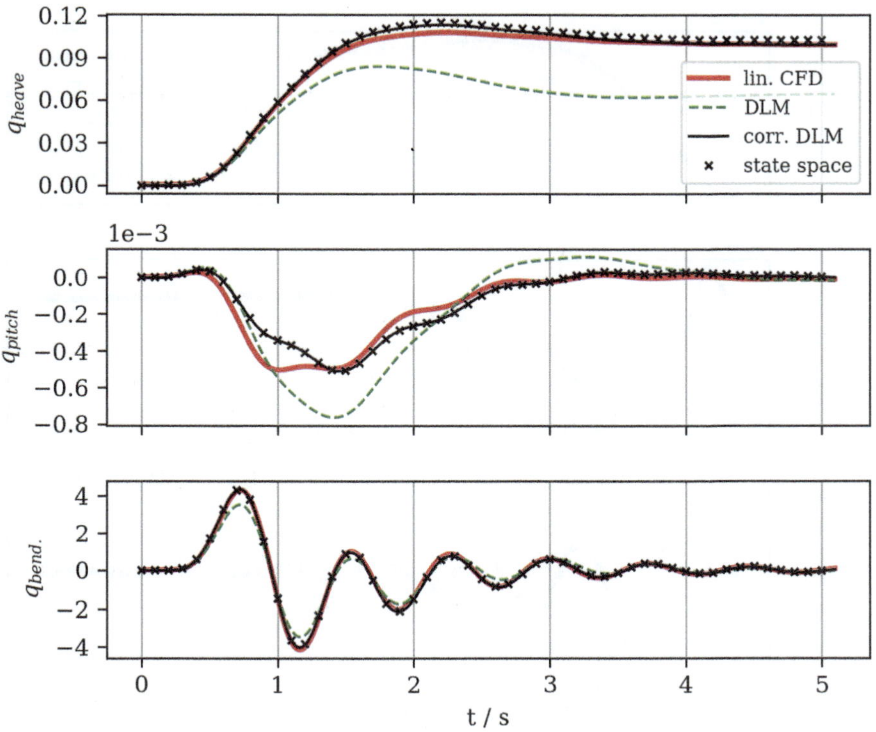

Fig. 8 Aeroelastic response of the CRM due to 1-cos gust encounter as displacements in modal coordinates for a gust length of 210 m

Overall, the time domain responses of the state space model provide confirmation in the RFA using the Vector Fitting Method. The excellent agreement with the transformed frequency responses makes this approach feasible for an aeroelastic reduced order model in the time domain.

4 Conclusion

This work presents an aeroelastic reduced order model for gust encounter simulations in the time domain which combines aeroelastic coupling in the frequency domain with an AIC-based correction method and a state space representation of the aeroelastic transfer functions. The resulting state space model allows for an efficient evaluation of aeroelastic responses due to signals on specified inputs comprising rigid body motions, elastic motions, control surface deflections and gust encounters. The aeroelastic ROM is demonstrated for the CRM configuration with 1-cos

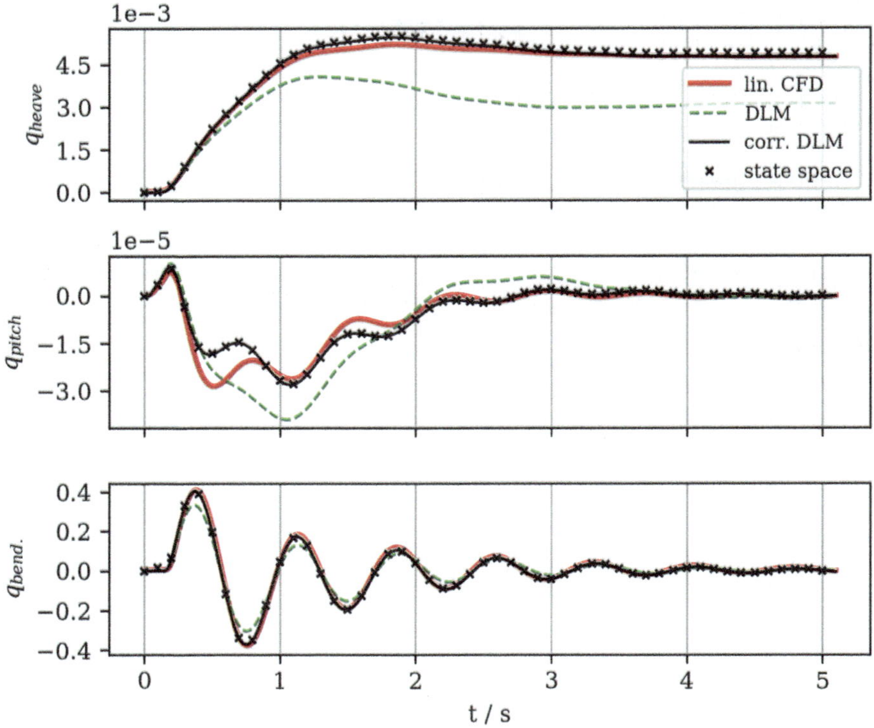

Fig. 9 Aeroelastic response of the CRM due to 1-cos gust encounter as displacements in modal coordinates for a gust length of 10 m

gust encounters. The obtained aeroelastic response in generalized forces and displacements shows an overall improvement of the DLM-based response in comparison with the time-linearized RANS-based response. The improvement results from the proposed correction method which allows to modify the AIC matrix in order to reproduce quasi-steady results from higher-order methods like time-linearized RANS solutions. Moreover, the approach of correcting the Taylor series of the AIC matrix extends the quasi-steady correction to higher frequencies. In addition, the correction is able to account for multiple quasi-steady pressure responses to different harmonic motions. Indeed, the presented aerodynamic responses of the Aerostabil wing show the necessity of considering multiple pressure responses in the correction for the application to aeroelastic models. The main limitation of the proposed method is the small improved frequency bandwidth which restricts the correction to low frequency excitation. Increasing the correction order from the quasi-steady correction to higher order corrections can further improve the results but, in general, the frequency bandwidth remains small due to the limited radius of convergence of the underlying Taylor series expansion. Therefore, the proposed correction method needs extension for higher frequencies in order to increase the applicability of the presented reduced order model. For the presented gust encounter simulations, the low frequency content is more significant resulting in improved aeroelastic responses over the DLM-based solutions.

Acknowledgements The funding of partial activities through the Federal Ministry for Economic Affairs and Climate Action under contracts FKZ 20A1504A in the framework of the air transport research program is gratefully acknowledged.

References

1. Albano, E., Rodden, W.P.: A doublet-lattice method for calculating lift distributions on oscillating surfaces in subsonic flows. AIAA J. **7**(2), 279–285 (1969)
2. Brendes, G., Voss, R.: Transonic flutter calculations for a low wing transport aircraft using transonic doublet lattice method. In: International Forum on Aeroelasticity and Structural Dynamics (IFASD) 2001, Madrid, Spain, 5–7 June (2001)
3. Brink-Spalink, J., Bruns, J.M.: Correction of unsteady aerodynamic influence coefficients using experimental or CFD data. In: International Forum on Aeroelasticity and Structural Dynamics (IFASD) 2001, Madrid, Spain, 5–7 June (2001)
4. Quero-Martin, D.: An aeroelastic reduced order model for dynamic response prediction to gust encounters. Ph.D. thesis, TU Berlin (2017)
5. Palacios, R., Climent, H., Karlsson, A., Winzell, B.: Assessment of strategies for correcting linear unsteady aerodynamics using CFD or experimental results, pp. 209–224. Progress in Computational Flow-Structure Interaction. Springer (2003)
6. Thormann, R., Dimitrov, D.: Correction of aerodynamic influence matrices for transonic flow. CEAS Aeronaut. J. **5**(4), 435–446 (2014)
7. Vassberg, J., Dehaan, M., Rivers, M., Wahls, R.: Development of a common research model for applied CFD validation studies. In: 26th AIAA Applied Aerodynamics Conference, Honolulu, Hawaii, USA, 18–21 August (2008)
8. Kaiser, C., Quero, D., Nitzsche, J.: Quantification of nonlinear effects in gust load prediction. In: International Forum on Aeroelaticity and Structural and Dynamics (IFASD) 2019, Savannah, Georgia, USA, 9–13 June (2019)
9. Gustavsen, B., Semlyen, A.: Rational approximation of frequency domain responses by vector fitting. IEEE Trans. Power Deliv. **14**(3), 1052–1061 (1999)
10. Friedewald, D., Thormann, R., Kaiser, C., Nitzsche, J.: Quasi-steady doublet-lattice correction for aerodynamic gust response prediction in attached and separated transonic flow. CEAS Aeronaut. J. **9**(1), 53–66 (2017)
11. Kaiser, C., Friedewald, D., Nitzsche, J.: Comparison of nonlinear CFD with time-linearized CFD and CFD-correced DLM for gust encounter simulations. In: International Forum on Aeroelaticity and Structural and Dynamics (IFASD) 2017, Como, Italy, 25–28 June (2017)
12. Dietz, G., Schewe, G., Kiessling, F., Sinapius, M.: Limit-cycle-oscillations experiments at a transport aircraft wing model. In: International Forum on Aeroelasticity and Structural Dynamics (IFASD) 2003, Amsterdam, Netherlands, 4–6 June (2003)
13. Schwamborn, D., Gerhold, T., Heinrich, R.: The DLR TAU-code: recent applications in research and industry. In: European Conference on Computational Fluid Dynamics (ECCOMAS) (2006)
14. Hirt, C., Amsden, A.A., Cook, J.L.: An arbitrary Lagrangian-Eulerian computing method for all flow speeds. J. Comput. Phys. **3**(14), 227–253 (1974)
15. Spalart, P., Allmaras, S.: A one-equation turbulence model for aerodynamic flows. In: 30th Aerospace Sciences Meeting and Exhibit, AIAA-92-0439 (1992)

16. Thormann, R., Widhalm, M.: Linear-frequency-domain predictions of dynamic-response data for viscous transonic flows. AIAA J. **51**(11) (2013)
17. Kaiser, C., Thormann, R., Dimitrov, D., Nitzsche, J.: Time-linearized analysis of motion-induced and gust-induced airloads with the DLR TAU code. In: Deutscher Luft- und Raumfahrtkongress 2015, Rostock, Germany, Deutsche Gesellschaft für Luft- und Raumfahrt - Lilienthal-Oberth e.V., Bonn (2015)
18. 4th AIAA Drag Prediction Workshop (2017). https://dpw.larc.nasa.gov/DPW4/DLR/SolarGrids
19. Klimmek, T.: Parametric set-up of a structural model for FERMAT configuration for aeroelastic and loads analysis. J. Aeroelasticity Struct. Dyn. **2**(3), 31–49 (2014)

System Identification from Virtual Flight Test Data for a Flexible Transport Aircraft

Mathias S. Roeser and Wulf Mönnich

Abstract This chapter describes the latest achievements from system identification of flexible aircraft using virtual flight test data. The goal is to arrive at a model that is real-time capable and can be used in a flight simulator. Virtual flight tests were performed using high-fidelity multidisciplinary CFD/CSM methods. For that a new method for input design was applied. A description of the flexible aircraft model is included and this model structure is adopted for the system identification process. The time domain system identification for the flexible aircraft was split in two parts: first the rigid-body dynamics were identified and then a new process to estimate the structural dynamics was developed and applied. Results for both parts will be shown.

Keywords Aircraft system identification · Maneuver design · Virtual flight testing · Multi-input signals · Wavelet packets

1 Introduction

The development of modern aircraft shows two branches demanding closer evaluations of the structural dynamics: on one side highly flexible wing designs as a consequence of structurally optimized aircraft configurations and on the other side very large aircraft that result also in very flexible wing structures. One of the effects of such designs is that the frequencies of the lower-frequency structural modes may be in the vicinity of the frequencies of the rigid-body dynamics. This may affect the stability and control and furthermore the handling qualities of these types of aircraft.

M. S. Roeser (✉) · W. Mönnich
German Aerospace Center, Institute of Flight Systems, Lilienthalplatz 7, 38108 Braunschweig, Germany
e-mail: Mathias.Roeser@deutscheaircraft.com

W. Mönnich
e-mail: wulf.moennich@dlr.de

R. Heinrich (ed.), *Advanced Aircraft Understanding via the Virtual Aircraft Model*, Notes on Numerical Fluid Mechanics and Multidisciplinary Design 155,
https://doi.org/10.1007/978-3-031-69425-7_5

The impact on the handling qualities (HQ) is a challenging and relatively new field of research. For HQ investigations in a simulator accurate, real-time-capable, flexible aircraft models have to be developed.

With increasing accuracy of multidisciplinary flight simulation tools the substitution of real flight tests by numerical simulations becomes a field of interest to reduce the cost of real flight test campaigns for system identification. These multidisciplinary flight simulation tools consolidate a high-fidelity computational fluid dynamics (CFD) code for the determination of the aerodynamic forces and moments, a computational structural mechanics (CSM) code to cover the elastic motions and a flight dynamics module. Additionally, high-fidelity CFD/CSM methods can assist in the evaluation of critical configurations and operating points and may become essential for decisions about design changes. However, the high computational demand of these high-fidelity CFD/CSM simulations still makes it impossible to use this tool directly in real-time for flight mechanical and handling qualities evaluations or for controller design.

Virtual flight tests (VFT) have several advantages over actual flight tests with regard to system identification, e.g.:

- The aircraft's mass, its inertia tensor, the position of the center of gravity and the thrust are perfectly known.
- The CFD/CSM simulation data used as virtual flight test (VFT) data contain no measurement errors and many more quantities can be "measured" directly, e.g. the aerodynamic forces and moments or the sideslip angle. For the structural dynamics even the generalized coordinates and their first- and second-order derivatives can be provided.
- Control surface deflections and rates that are not possible in actual flight can be simulated, e.g. symmetrical aileron deflections and high rates.

However, there are also constraints:

- The control deflections of the identification maneuvers should be kept in physically reasonable ranges and also within the simulation capabilities of the CFD code.
- In VicToria and VitAM it was not yet possible to get reasonable results for the deflections of spoilers.
- The time and costs associated with the generation of the VFT data: the duration of the system identification maneuver that allows the identification of a complete aerodynamic model valid in the vicinity of the trim point should not exceed 10 s. This is a very short duration considering that the typical single-surface-excitation maneuvers for an actual flight test last anywhere from 15 s to more than one minute.

Due to the limit on the duration a maneuver is required where several control surfaces are deflected. A new method to design such a maneuver is described in Sect. 2.

A short Sect. 3 regarding the VFT data is followed by the description of the mathematical aircraft model used for the identification (Sect. 4). The model accounts for all relevant rigid-body parameters, structural dynamics parameters and those parameters which describe the interaction between the structural and the rigid-body dynamics.

To extract a real-time-capable model from the VFT data classical system identification and parameter estimation techniques are pursued. However, depending on the number of structural modes to be included in the model, the number of parameters to be estimated can increase immensely. Therefore, a new system identification approach is required.

An automatic process for the identification of a complete aircraft model is presented in Sect. 5. It includes a categorization of each structural mode shape and several automatically performed steps allowing an intuitive tracking of any intermediate results. The results obtained from multi-axis system identification maneuvers are shown and they illustrate the advances in the virtual flight testing area.

2 Input Design for Virtual Flight Tests

Input design for system identification maneuvers still strongly relies on classical inputs, e.g. doublets, pulses or multi-step inputs, specially if those maneuvers are to be flown manually by pilots. Some effort was given to adapt those classical inputs, even if manually flown, for instance by adjusting the time step of each individual step input within one maneuver input, to obtain a better distribution of the frequency content of the signals (Chap. 2.3 of Ref. [3]) or to optimize the inputs with regard to the information content [4].

With increasing automation new maneuvers can be performed, for instance the combined excitation of several control surfaces at the same time [5, 6, 8, 9]. These new maneuvers considerably increase the information content in the data, however might not always be applicable, especially in such a research topic as covered by this book, due to the complex interpretability of the results using time domain methods. If unexpected effects, for instance aerodynamic effects derived from CFD/CSM computation, arise during the simulations performed within this research, it would become considerably difficult to identify the source, for instance if it is a CFD/CSM solver issue, a model postulation deficiency or any other effect. Therefore, an interpretable input design based on classical designs is pursued to support the human-based model structure determination.

An additional challenge for VFT simulations is to combine a high information content of the VFT data with a short maneuver time as the simulations to be performed with high-fidelity CFD methods are very time consuming. Within the VitAM project, a very short maneuver of less than 5 s was pursued to get information about the longitudinal dynamics of an Airbus A320 aircraft including the elevator, δ_e, horizontal stabilizer, i_{HT}, and symmetric aileron, $\delta_{a,sym}$, control surface effectivenesses. For comparison, classically a 10–15 s maneuver is used to identify the short period dynamics using only one control surface, for instance the elevator. To comply with such a short-time maneuver a combination of inputs was chosen that allowed the separation in time for each control surface resulting in a decorrelation of the inputs and a good interpretability by the user. With a preliminary derivative model as base for the input design, the combination of inputs that showed the best predicted results for the identification was used as candidate maneuver.

Virtual flight tests additionally allow to specify maneuvers without the need to take into account the actuator dynamics of the real flight test aircraft. This is a great advantage in relation to real flight test data, where often the dynamics of the control surface deflections have to be considered. Nevertheless, the control surface displacements should be kept within physically realistic limits and should not result in any instability of the CFD code e.g. due to large step-like deflections.

The method described in Ref. [10] allowed the inclusion of all constraints for the input to be simulated. The use of smooth wavelet functions with no sharp edges avoided the inclusion of undesired high-frequency content, as it would be the case for step-like multi-step inputs. The parametric design method allowed the combination of several inputs. Besides the VitAM maneuver another maneuver, also exciting the lateral-directional motion by aileron, δ_a, and rudder, δ_r, deflections, was designed for the VicToria project. In addition, an independent proof-of-match maneuver (PoM) was designed to verify that the parameters identified from the VitAM and the VicToria maneuvers also work for a different maneuver, thereby validating the identified model. The input time histories of all three maneuvers are shown in Fig. 1.

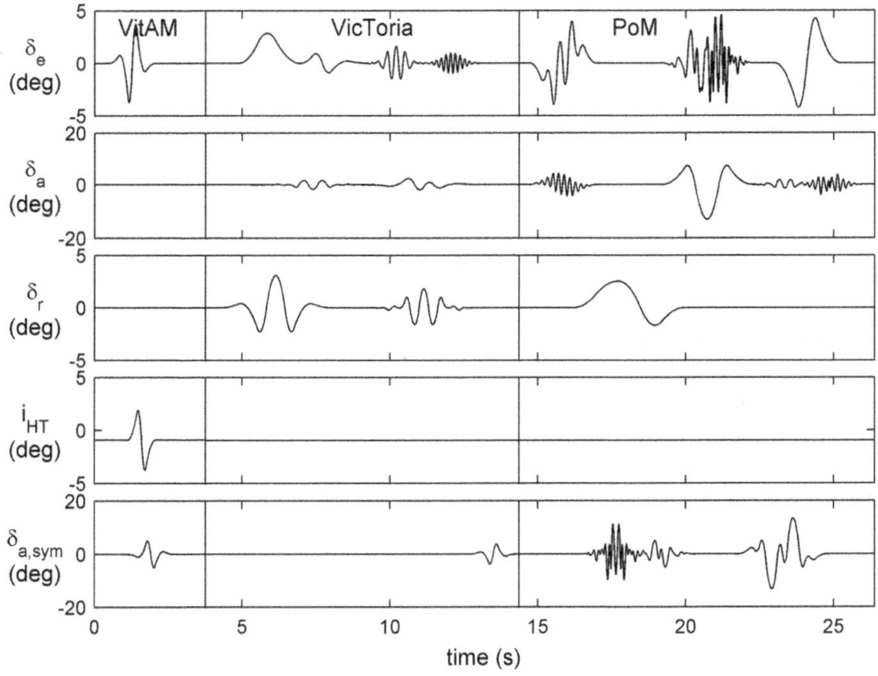

Fig. 1 Control surface inputs

3 VFT Data

In addition to the VFT data directly provided by the CFD/CSM simulations (cf. Table 1 of Chap. 2), the accelerations that would be measured by an error-free accelerometer at the center of gravity were calculated from the VFT data as additional "measurement" channels

$$\begin{aligned} a_x &= \dot{u} - rv + qw + g \sin\theta \\ a_y &= \dot{v} - pw + ru - g \sin\phi \cos\theta \\ a_z &= \dot{w} - qu + pv - g \cos\phi \cos\theta \end{aligned} \tag{1}$$

4 Flexible Aircraft Model

4.1 Equations of Motion for the System Identification of a Flexible Aircraft

The model presented in this section is based on the work of Waszak and Schmidt [12]. The derivation of the equations of motion and the inertial decoupling between the rigid-body and the structural dynamics is obtained through the use of the mean axes reference system and the following assumptions: 1. structural deformation is assumed to be small and described by a set of eigenmodes; 2. moments and products of inertia do not vary with the deformation; 3. deformation and deformation rates are collinear. The first assumption permits writing the structural displacement of a point of the aircraft's structure as an infinite sum of contributions from its normal modes.

$$\mathbf{d}(x, y, z, t) = \sum_{k=1}^{\infty} \mathbf{\Phi}(x, y, z)\eta_k(t) \tag{2}$$

In Eq. (2), the vector $\mathbf{\Phi}$ represents the shape of the structural modes and has dimension of length, whereas $\eta_k(t)$ is a scalar function of time and represents the generalized displacement of the k-th mode. For real applications in simulation or system identification, the sum in Eq. (2) must be truncated to n_F modes at some point, which means that some normal modes should be retained and the remaining ones should be discarded. If the interest lies on the influence of the flexibility on the flight dynamics, the most straightforward way to make this selection is to retain the structural modes with lowest natural frequencies. These structural modes have the largest influence on the dynamics of the rigid-body modes. Note that in our case the VFT data were already generated using the modal approach, retaining the 21 structural modes with lowest natural frequencies.

From the kinetic and potential energy expressions (including the energy terms due to structural deformation), the application of Lagrange's equation leads to the following set of equations of motion

$$\begin{aligned}
&\dot{u} - rv + qw + g\sin\theta = X/m_{ac} \\
&\dot{v} - pw + ru - g\sin\phi\cos\theta = Y/m_{ac} \\
&\dot{w} - qu + pv - g\cos\phi\cos\theta = Z/m_{ac} \\
&I_{xx}\,\dot{p} - (I_{xy}\,\dot{q} + I_{xz}\,\dot{r}) + (I_{zz} - I_{yy})qr + (I_{xy}\,r - I_{xz}\,q)p + (r^2 - q^2)\,I_{yz} = \mathcal{L} \\
&I_{yy}\,\dot{q} - (I_{xy}\,\dot{p} + I_{yz}\,\dot{r}) + (I_{xx} - I_{zz})pr + (I_{yz}\,p - I_{xy}\,r)q + (p^2 - r^2)\,I_{xz} = \mathcal{M} \\
&I_{zz}\,\dot{r} - (I_{xz}\,\dot{p} + I_{yz}\,\dot{q}) + (I_{yy} - I_{xx})pq + (I_{xz}\,q - I_{yz}\,p)r + (q^2 - p^2)\,I_{xy} = \mathcal{N} \\
&\ddot{\eta}_k + 2\zeta_k\,\omega_k\,\dot{\eta}_k + \omega_k^2\eta_k = Q_k/m_k \qquad k = 1, ..., n_F
\end{aligned} \tag{3}$$

The equations of motion are completed by the kinematic relations for the rotation and altitude change, which are identical to the classic rigid-body motion

$$\begin{aligned}
\dot{\phi} &= p + q\sin\phi\tan\theta + r\cos\phi\tan\theta \\
\dot{\theta} &= q\cos\phi - r\sin\phi \\
\dot{\psi} &= (q\sin\phi + r\cos\phi)\sec\theta \\
\dot{h} &= u\sin\theta - v\sin\phi\cos\theta - w\cos\phi\cos\theta
\end{aligned} \tag{4}$$

The first six equations in Eq. (3) are written with regard to the mean axes reference system and are formally equivalent to the classical equations for rigid-body dynamics. The last expression models the structural response in terms of the modal deflections (also called generalized displacement) η_k. The model has a total of $(6 + n_F)$ degrees of freedom, where n_F is the number of structural modes retained. The DOFs correspond to three translations, three rotations, and the structural modal displacements. Due to the assumption of constant inertia and the use of the mean axes system, the rigid-body and the structural motion are coupled solely through the aerodynamic forces and moments on the right-hand side.

In Eq. (3) the state variables are the body-fixed airspeed components u, v and w, the body-fixed rotational rate components p (roll rate), q (pitch rate) and r (yaw rate), the attitude angles ϕ (roll angle), θ (pitch angle) and ψ (yaw angle), the altitude h and the generalized coordinates of the structural modes η_k and their time derivatives $\dot{\eta}_k$. X, Y, and Z are the external forces acting on the respective axes. $\mathcal{L}$, $\mathcal{M}$ and $\mathcal{N}$ are the external moments about the center of gravity. The aircraft mass and moments and cross products of inertia are given by m_{ac}, I_{xx}, I_{yy}, I_{zz}, I_{xy}, I_{xz} and I_{yz}. Q_k represents the generalized force acting on the k-th structural mode and has aerodynamic source. The modal mass, in-vacuo natural frequency and damping ratio are, respectively m_k, ω_k and ζ_k. The VFT data were generated with the mass-normalized version of this equation, i.e., $m_k = 1$ kg, and the structural damping was assumed to be negligible, i.e., $\zeta_k = 0$, for all k.

The aerodynamic forces and moments are assumed to be composed of a superposition of loads due to rigid-body motion and control variables (subscript RB) and due to flexible response (subscript $Flex$)

$$\begin{aligned} X &= X_{RB} + X_{Flex} & \mathcal{L} &= \mathcal{L}_{RB} + \mathcal{L}_{Flex} \\ Y &= Y_{RB} + Y_{Flex} & \mathcal{M} &= \mathcal{M}_{RB} + \mathcal{M}_{Flex} \\ Z &= Z_{RB} + Z_{Flex} & \mathcal{N} &= \mathcal{N}_{RB} + \mathcal{N}_{Flex} \end{aligned} \tag{5}$$

The same approach is adopted for the aerodynamic generalized loads acting on the structural degrees of freedom

$$Q_k = Q_{k,RB} + Q_{k,Flex} \qquad k = 1, ..., n_F \tag{6}$$

To model the influence of rigid-body and flexible variables on the aerodynamic forces, moments, and generalized loads, a stability and control derivatives approach is adopted.

4.2 *Aerodynamic Model*

To express the forces and moments resulting from rigid-body motion and control deflections, a conventional stability and control derivative model is used. The forces and moments are obtained from the corresponding coefficients through the definitions

$$\begin{aligned} X_{RB} &= \bar{q} S\, C_{X,RB} & \mathcal{L}_{RB} &= \bar{q} S s\, C_{l,RB} \\ Y_{RB} &= \bar{q} S\, C_{Y,RB} & \mathcal{M}_{RB} &= \bar{q} S \bar{c}\, C_{m,RB} \\ Z_{RB} &= \bar{q} S\, C_{Z,RB} & \mathcal{N}_{RB} &= \bar{q} S s\, C_{n,RB} \end{aligned} \tag{7}$$

where $\bar{q}$ is the dynamic pressure, S is the wing reference area, $\bar{c}$ is the mean aerodynamic chord, and s is the semispan of the wing. The dynamic pressure is calculated from the air density ρ (which is in our case constant and the same as in the CFD simulations) and the true airspeed V_{TAS}

$$\bar{q} = \frac{\rho}{2} V_{\text{TAS}}^2 \tag{8}$$

with $V_{\text{TAS}} = \sqrt{u^2 + v^2 + w^2}$.

The model contains control surface deflections of the elevator, δ_e, the horizontal tail, i_{HT}, the left and right ailerons, $\delta_{a,l}$, $\delta_{a,r}$, and the rudder, δ_r. The classical asymmetric aileron deflection δ_a and the symmetric deflection $\delta_{a,sym}$ are computed from the left and right deflections

$$\delta_a = \frac{1}{2}(\delta_{a,r} - \delta_{a,l}) \qquad \delta_{a,sym} = \frac{1}{2}(\delta_{a,r} + \delta_{a,l}) \tag{9}$$

The equations for the longitudinal motion are based on the two-point model described in Refs. [3, 7]. This model separates the wing and horizontal tail influences and allows to account for the downwash lag effect from the wing to the horizontal tail.

In the following equations of the aerodynamic model the parameters to be estimated are boldfaced.

The lift coefficient is separated into a wing-body (WB) component and a horizontal tail (HT) component. The drag coefficient is calculated via the simple polar equation. The pitching moment coefficient about a body-fixed aerodynamic reference point (ARP) is also separated into a wing-body component and a horizontal tail component calculated via the body-axis force coefficients $C_{X,HT}$ and $C_{Z,HT}$ and the respective lever arms x_{HT} and z_{HT}.

$$
\begin{aligned}
C^e_{L,RB} &= C^e_{L,WB} + C^e_{L,HT}\frac{S_{HT}}{S}\cos(\alpha_{dyn} - \varepsilon_{HT}) \\
C^e_{D,RB} &= \boldsymbol{C_{D0}} + \boldsymbol{k}C^e_{L,RB} + \frac{(C^e_{L,RB})^2}{e\pi\Lambda} - C^e_{L,HT}\frac{S_{HT}}{S}\sin(\alpha_{dyn} - \varepsilon_{HT}) \\
C^e_{m,RB} &= C^e_{m,WB} + C_{Z,HT}\frac{S_{HT}}{S}\frac{x_{HT}}{\bar{c}} - C_{X,HT}\frac{S_{HT}}{S}\frac{z_{HT}}{\bar{c}}
\end{aligned}
\tag{10}
$$

where S_H is the reference area of the horizontal tail, α_{dyn} is the change in the angle of attack of the horizontal tail due to a pitching motion of the aircraft, ε_{HT} is the downwash angle at the horizontal tail and Λ is the aspect ratio of the wing. The superscript "e" indicates that the coefficients are expressed in the intermediate system as defined in the ISO 1151-1 standard [1]. The transformation from the intermediate to the body-fixed system involves just the angle of attack α and not the sideslip angle β (cf. Eq. (16)).

The separated influences of wing and tail for the longitudinal motion are calculated by

$$
\begin{aligned}
C^e_{L,WB} &= \boldsymbol{C_{L0}} + \boldsymbol{C_{L\alpha,WB}}\alpha \\
C^e_{L,HT} &= \boldsymbol{C_{L\alpha,HT}}\alpha_{HT} + \boldsymbol{C_{L\delta_e,HT}}\delta_e \\
C^e_{m,WB} &= \boldsymbol{C_{m0,WB}} + \boldsymbol{C_{mq,WB}}q^* + \boldsymbol{C_{m\delta_{a,sym}}}\delta_{a,sym} \\
C_{X,HT} &= C^e_{L,HT}\sin(\alpha_{HT} - i_{HT}) \\
C_{Z,HT} &= -C^e_{L,HT}\cos(\alpha_{HT} - i_{HT})
\end{aligned}
\tag{11}
$$

where α is the angle of attack, α_{HT} is the angle of attack of the horizontal tail and q^* is the nondimensional pitch rate.

$$\begin{aligned}\alpha &= \tan^{-1}\frac{w}{u}\\ \alpha_{HT} &= \alpha + i_{HT} - \varepsilon_{HT} + \alpha_{dyn}\\ q^* &= \frac{q\bar{c}}{V_{\mathrm{TAS}}}\end{aligned} \tag{12}$$

The downwash angle at the horizontal tail ε_{HT} describes the influence of the wing on the tail

$$\varepsilon_{HT} = \varepsilon_{0,HT} + \frac{\partial \boldsymbol{\varepsilon}}{\partial \boldsymbol{\alpha}}\alpha(t-\tau) \tag{13}$$

where $\varepsilon_{0,HT}$ is the downwash angle for $\alpha = 0$, calculated by $\varepsilon_{0,HT} = \frac{C_{L0}}{C_{L\alpha,WB}}\frac{\partial\varepsilon}{\partial\alpha}$, and τ is the time delay $\tau = \frac{x_{HT}}{V_{\mathrm{TAS}}}$ to consider the transport lag of a change in the flow from the wing to the tail with x_{HT} being the x-position of the aerodynamic center of the wing relative to the one of the horizontal tail. α_{dyn} is the change in the angle of attack of the horizontal tail due to a pitching motion of the aircraft and can be approximated by

$$\alpha_{dyn} = \tan^{-1}\left(q\frac{x'_{HT}}{V_{\mathrm{TAS}}}\right) \tag{14}$$

where x'_{HT} is the x-position of the center of gravity relative to the aerodynamic center of the horizontal tail.

The coefficients of the aerodynamic forces and moments for the lateral-directional dynamics ($C^e_{Y,RB}$, $C^e_{l,RB}$, $C^e_{n,RB}$) are derived by simple Taylor series expansions

$$\begin{aligned}C^e_{Y,RB} &= \boldsymbol{C_{Y\beta}}\beta + \boldsymbol{C_{Yp}}p^* + \boldsymbol{C_{Yr}}r^* + \boldsymbol{C_{Y\delta_r}}\delta_r\\ C^e_{l,RB} &= (\boldsymbol{C_{l\beta}} + \boldsymbol{C_{l\beta_\alpha}}\alpha)\beta + (\boldsymbol{C_{lp}} + \boldsymbol{C_{lp_\alpha}}\alpha)p^* + (\boldsymbol{C_{lr}} + \boldsymbol{C_{lr_\alpha}}\alpha)r^*\\ &\quad + \boldsymbol{C_{l\delta_a}}\delta_a + \boldsymbol{C_{l\delta_r}}\delta_r\\ C^e_{n,RB} &= (\boldsymbol{C_{n\beta}} + \boldsymbol{C_{n\beta_\alpha}}\alpha)\beta + (\boldsymbol{C_{np}} + \boldsymbol{C_{np_\alpha}}\alpha)p^* + (\boldsymbol{C_{nr}} + \boldsymbol{C_{nr_\alpha}}\alpha)r^*\\ &\quad + \boldsymbol{C_{n\delta_a}}\delta_a + \boldsymbol{C_{n\delta_r}}\delta_r\end{aligned} \tag{15}$$

$$\text{with}\quad \beta = \sin^{-1}\frac{v}{V_{\mathrm{TAS}}} \qquad p^* = \frac{ps}{V_{\mathrm{TAS}}} \qquad r^* = \frac{rs}{V_{\mathrm{TAS}}}$$

where β is the sideslip angle and p^* and r^* are the nondimensional roll and yaw rates.

The body-axis force coefficients, $C_{X,RB}$, $C_{Y,RB}$ and $C_{Z,RB}$, and the body-axis moment coefficients about the center of gravity, $C_{l,RB}$, $C_{m,RB}$ and $C_{n,RB}$, are finally derived by

$$
\begin{aligned}
C_{X,RB} &= -C_D^e \cos\alpha + C_L^e \sin\alpha \\
C_{Y,RB} &= C_{Y,RB}^e \\
C_{Z,RB} &= -C_L^e \cos\alpha - C_D^e \sin\alpha \\
C_{l,RB} &= C_{l,RB}^e \cos\alpha - C_{n,RB}^e \sin\alpha + \frac{1}{s}(C_{Z,RB}\, y_{ARP,CG} - C_{Y,RB}\, z_{ARP,CG}) \\
C_{m,RB} &= C_{m,RB}^e + \frac{1}{\bar{c}}(-C_{Z,RB}\, x_{ARP,CG} + C_{X,RB}\, z_{ARP,CG}) \\
C_{n,RB} &= C_{l,RB}^e \sin\alpha + C_{n,RB}^e \cos\alpha + \frac{1}{s}(C_{Y,RB}\, x_{ARP,CG} - C_{X,RB}\, y_{ARP,CG})
\end{aligned}
\tag{16}
$$

where $[x_{ARP,CG}\ y_{ARP,CG}\ z_{ARP,CG}]$ is the position of the aerodynamic reference point relative to the center of gravity in body-fixed coordinates.

When dealing with flexible aircraft, the aerodynamic force and moment coefficients that drive the rigid-body dynamics receive contributions of the structural deformation and its changes. In general, the shapes of the structural eigenmodes correspond to a combination of bending and torsion of the lifting surfaces and the fuselage and so, the generated forces and moments are influenced by modal deflections and their rates.

Equation (17) shows how the influence of the structural modes on the force and moment coefficients is calculated

$$
\begin{aligned}
C_{X,Flex} &= \sum_{k=1}^{n_F} C_{X\eta_k}\eta_k + \sum_{k=1}^{n_F} C_{X\dot{\eta}_k}\frac{\dot{\eta}_k \bar{c}}{2V_{\text{TAS}}} \\
C_{Y,Flex} &= \sum_{k=1}^{n_F} C_{Y\eta_k}\eta_k + \sum_{k=1}^{n_F} C_{Y\dot{\eta}_k}\frac{\dot{\eta}_k \bar{c}}{2V_{\text{TAS}}} \\
C_{Z,Flex} &= \sum_{k=1}^{n_F} C_{Z\eta_k}\eta_k + \sum_{k=1}^{n_F} C_{Z\dot{\eta}_k}\frac{\dot{\eta}_k \bar{c}}{2V_{\text{TAS}}} \\
C_{l,Flex} &= \sum_{k=1}^{n_F} C_{l\eta_k}\eta_k + \sum_{k=1}^{n_F} C_{l\dot{\eta}_k}\frac{\dot{\eta}_k \bar{c}}{2V_{\text{TAS}}} \\
C_{m,Flex} &= \sum_{k=1}^{n_F} C_{m\eta_k}\eta_k + \sum_{k=1}^{n_F} C_{m\dot{\eta}_k}\frac{\dot{\eta}_k \bar{c}}{2V_{\text{TAS}}} \\
C_{n,Flex} &= \sum_{k=1}^{n_F} C_{n\eta_k}\eta_k + \sum_{k=1}^{n_F} C_{n\dot{\eta}_k}\frac{\dot{\eta}_k \bar{c}}{2V_{\text{TAS}}}
\end{aligned}
\tag{17}
$$

The feedback of the structural dynamics on the rigid body is given by the forces and moments according to the definition in Eq. (18)

$$
\begin{aligned}
X_{Flex} &= \bar{q}SC_{X,Flex} & \mathcal{L}_{Flex} &= \bar{q}SbC_{l,Flex} \\
Y_{Flex} &= \bar{q}SC_{Y,Flex} & \mathcal{M}_{Flex} &= \bar{q}S\bar{c}C_{m,Flex} \\
Z_{Flex} &= \bar{q}SC_{Z,Flex} & \mathcal{N}_{Flex} &= \bar{q}SbC_{n,Flex}
\end{aligned}
\tag{18}
$$

The second source of aerodynamic coupling corresponds to the parameters that represent the effects of rigid-body motion variables on the elastic degrees of freedom. The resulting expression for the influence of rigid-body motion and control surface deflections on the generalized loads is given by

$$
\begin{aligned}
Q_{k,RB} = \bar{q}S\bar{c}\Big(& \boldsymbol{C}_{\eta_k 0} + \boldsymbol{C}_{\eta_k \alpha}\alpha + \boldsymbol{C}_{\eta_k \beta}\beta + \boldsymbol{C}_{\eta_k p}p^* + \boldsymbol{C}_{\eta_k q}q^* + \boldsymbol{C}_{\eta_k r}r^* \\
& + \boldsymbol{C}_{\eta_k \delta_a}\delta_a + \boldsymbol{C}_{\eta_k \delta_{a,sym}}\delta_{a,sym} \\
& + \boldsymbol{C}_{\eta_k \delta_e}\delta_e + \boldsymbol{C}_{\eta_k \delta_r}\delta_r + \boldsymbol{C}_{\eta_k i_{HT}} i_{HT}\Big) \\
& k = 1, \ldots, n_F
\end{aligned}
\tag{19}
$$

Similarly, the influence of modal deflections and velocities on the generalized loads is given by

$$
Q_{k,Flex} = \bar{q}S\bar{c}\left(\sum_{j=1}^{n_F} \boldsymbol{C}_{\eta_k \eta_j}\eta_j + \sum_{j=1}^{n_F} \boldsymbol{C}_{\eta_k \dot{\eta}_j}\frac{\dot{\eta}_j \bar{c}}{2V_{\mathrm{TAS}}}\right) \qquad k = 1, \ldots, n_F \tag{20}
$$

Equations (19) and (20) show great similarity with the traditional approach of stability derivatives, widely used in rigid-body system identification. So, by using this approach, the extension from a rigid-body to a flexible modeling approach, in terms of computational implementation and use of available system identification software, is relatively straightforward.

5 Time Domain System Identification Using the Output Error Method

The VitAM and the VicToria maneuvers (Fig. 1) were used for the actual identification and the PoM maneuver was used for validation. All three maneuvers start from the same transonic trim point at Mach 0.78, h = 11278 m (37000 ft), steady, symmetric, horizontal flight in clean configuration.

In order to identify a flexible aircraft model from VFT data, a three-step process is proposed:

1. Identify a classic rigid-body model using Eqs. (10)–(15)
2. Additionally account for flexibility effects on the rigid-body model by using Eq. (17) taking the generalized coordinates and their rates from VFT data (as inputs)
3. Add elastic degrees of freedom using Eqs. (19) and (20).

Each of these steps is iterative.

5.1 Rigid-Body Dynamics Identification from VFT

The rigid-body dynamics were identified using the output error method for parameter estimation in the time domain [3], providing maximum likelihood estimates of the parameters.

For the identification of the rigid-body model initial values of the state variables and of the parameters of the aerodynamic model are required. In this case the initial values of the state variables can be taken directly from the VFT data. These values were kept constant during the identification runs except for u_0, w_0, θ_0 and h_0, which were estimated to obtain the best possible fit of the VFT data even with an imperfect aerodynamic model.

Weighted outputs, i.e., outputs whose matches have an influence on the cost function to be minimized during the identification, were the translational accelerations (a_x, a_y, a_z), the rotational accelerations ($\dot{p}$, $\dot{q}$, $\dot{r}$), the rotational rates (p, q, r), the attitude angles (ϕ, θ, ψ), true airspeed, angle of attack and sideslip angle (V_{TAS}, α, β) and the altitude (h).

The following procedure for the identification of a rigid-body model proved to be robust and effective.

1. For the first identification run there is no need to start with good initial values for the parameters. Actually almost all parameters of Eqs. (10) to (15) can be initialized to zero. It is sufficient to provide very rough, non-zero initial values for four parameters. For the current investigation $C_{L\alpha,WB} = 5$, $C_{L\alpha,HT} = 3$, $\frac{\partial \varepsilon}{\partial \alpha} = 0.8$ and $e = 0.8$ were used. In this first step the linear drag parameter k and the α-dependencies of the lateral-directional derivatives (i.e., $C_{l\beta_\alpha}$, C_{lp_α}, C_{lr_α}, $C_{n\beta_\alpha}$, C_{np_α}, C_{nr_α}) need to be kept fixed at zero. All other parameters can be included in the active set, i.e. they can be estimated.
2. Starting from the result of the previous step, also include the parameters that were kept fixed at zero in the identification. This will result in an improved cost function, but typically some of the parameters will have a high (i.e., greater than 5%) relative standard deviation (Cramér-Rao bounds) or the new set of parameters will not work well with the PoM maneuver. This is usually due to overparameterization. The goal is to find a minimum set of parameters that recovers mostly the cost function value of the identification with all parameters, but with small (i.e. less than 2%) relative standard deviations and an improved match of the PoM maneuver.

Finding the minimum set of parameters is an iterative process. Start with adding just one parameter to the result of step 1. Check that the cost function has been significantly improved (e.g., reduced by at least 50%), that the relative standard deviation of the parameters is small and that the match of the PoM maneuver is improved. Only if the answer is yes in all three cases keep the parameter in the active set. Add the next parameter and proceed as before until all parameters have been tested. Note that adding a new parameter may lead to a high relative standard deviation of another parameter. In such a case remove the parameter with the high relative standard deviation from the active set. In our case only the parameters k, C_{lp_α} and C_{np_α} have been retained.

3. Additionally account for flexibility effects on the rigid-body model by using Eqs. (17) with the generalized coordinates and their rates taken from VFT data (as inputs), applying the same iterative process as in step 2.
4. If more than one maneuver has been used for the identification, fix C_{L0}, C_{D0} and $C_{m0,WB}$ at the identified values and identify instead separate elevator and aileron offsets for each maneuver.

The finally obtained match of all relevant motion variables for all six degrees of freedom can be seen in Figs. 2, 3 and 4.

The two maneuvers used for the identification are matched very well. However, the proof-of-match maneuver shows clear deficiencies. Comparing these deficiencies

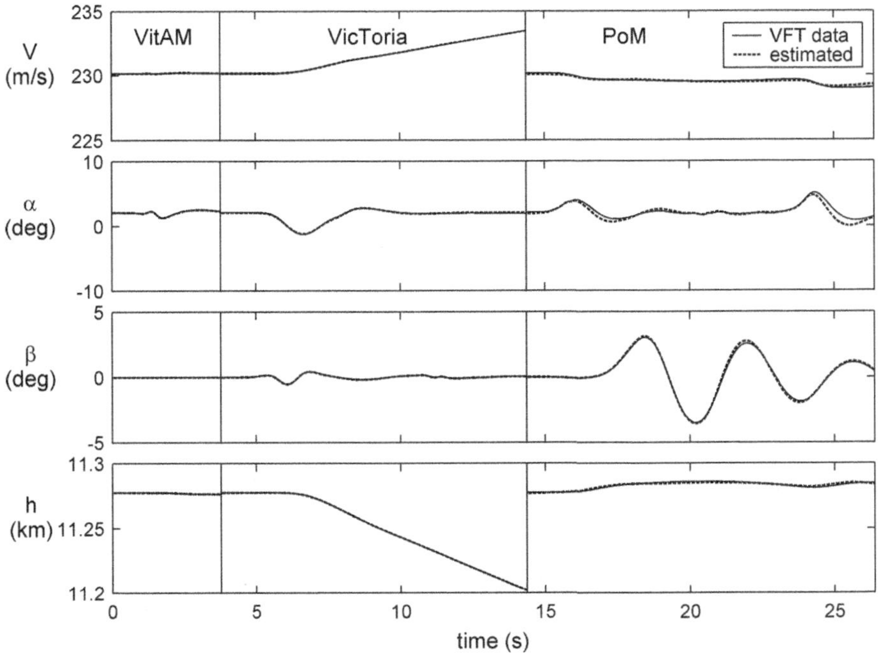

Fig. 2 Matches of flow variables and altitude

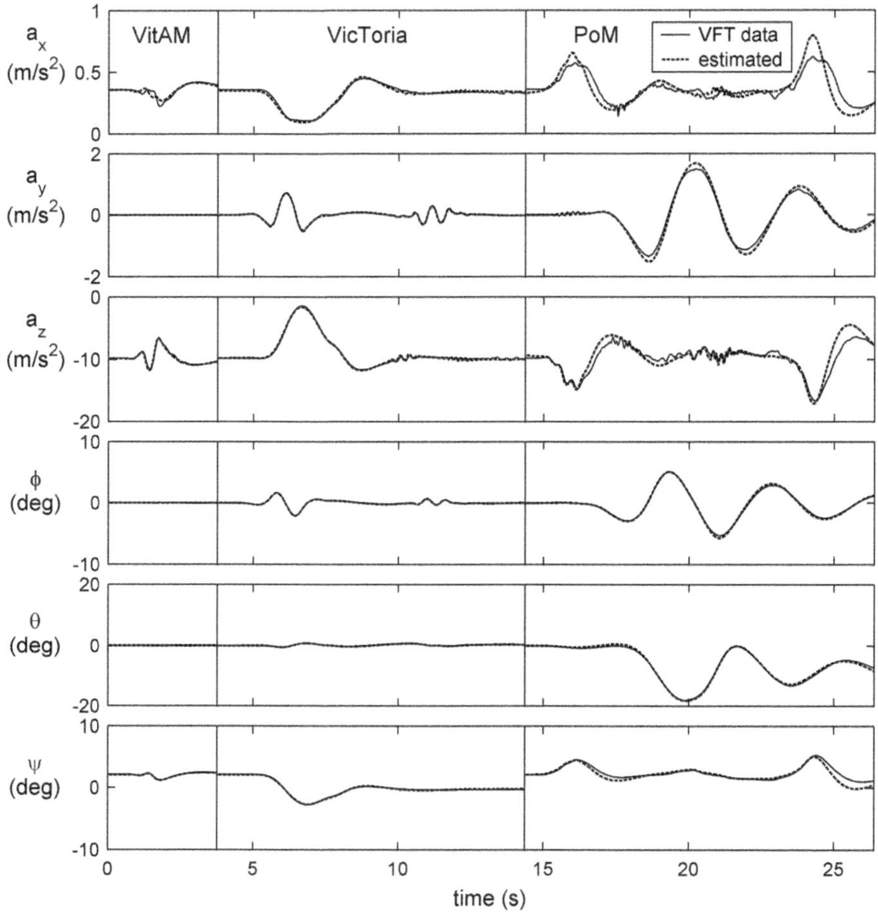

Fig. 3 Matches of translational accelerations and attitude angles

with the typical certification limits of Level D simulators [2] shows that only the vertical acceleration a_z is outside the limits. The maximum error is about 3.2 m/s^2, whereas the certification limit is just 1 m/s^2, so the validation clearly fails. The reasons can only be speculated. One possible reason could be unmodeled aerodynamic nonlinearities due to the transonic flight regime. To investigate this, further high-fidelity CFD/CSM simulations, e.g., at a lower Mach number, would be necessary.

Although the validation failed due to the large error in the vertical acceleration, it is remarkable that the rigid-body model which was identified from such short maneuvers matches a complex proof-of-match maneuver (mostly) inside the generalized Level D tolerances.

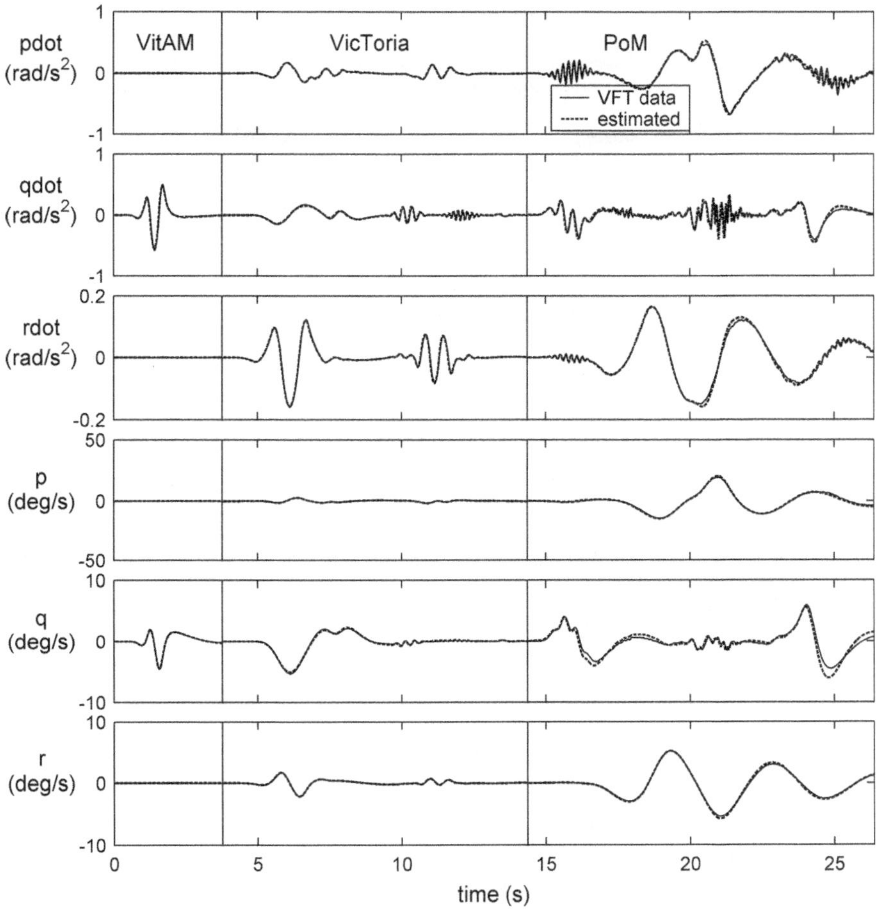

Fig. 4 Matches of rotational accelerations and rates

5.2 *A Process to Estimate Structural Dynamics Parameters for the Flexible Aircraft Model*

A more detailed description of the developed process can be found in Ref. [11].

The flexible aircraft model as described in Sect. 4 is an extension of the rigid-body aircraft model using a derivative formulation. It contains many additional parameters to describe the structural dynamics and the interaction of the structural dynamics with the rigid-body dynamics.

For the Airbus A320 a visualization of the first 21 structural modes (i.e., the ones that were used for the generation of the VFT data) is shown in Fig. 5.

It is directly comprehensible that the number of parameters needed to describe the structural dynamics increases significantly with the number of flexible modes

Fig. 5 Visualization of the first 21 structural modes of the A320 aircraft model

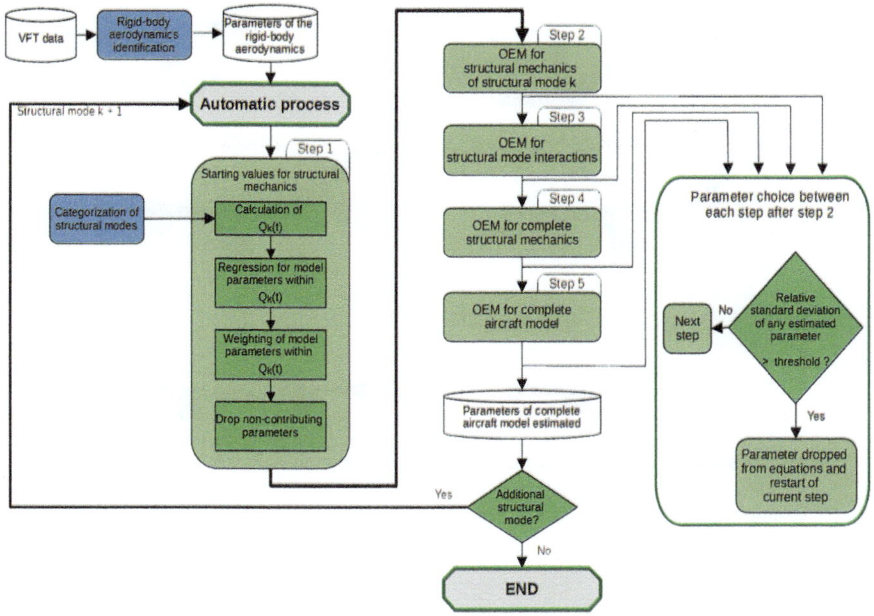

Fig. 6 Process for estimation of the complete aircraft model; blue blocks are user inputs, green blocks indicate automatic routines

to be included in the model. For instance, if the model formulation as described in Eqs. 17, 19 and 20 is used accounting for 21 modes, as it was the case in VitAM, 1365 parameters are available for estimation, which makes it almost impossible for a manual identification process, as it was the case for the rigid-body dynamics. Consequently a new strategy was used, see Fig. 6, and is briefly described in the following.

A. Categorization of Structural Modes

The structural modes can be classified according to the resulting aircraft structural deformation. Two types of mode shapes are defined: the longitudinal (symmetric) mode shapes and the lateral-directional (asymmetric) mode shapes. Each of these mode shapes is assumed to have different effects on the aircraft's aerodynamics and these assumptions are described in the following.

Longitudinal Mode Shape (Symmetric) The longitudinal modes are mode shapes that are symmetric to the aircraft's x-z-plane. The wing symmetric bending modes, horizontal tail symmetric modes and symmetric engine modes are the main contributors. For the estimation of the parameters of a symmetric mode, the longitudinal components α, q^*, δ_e, δ_{HT} and $\delta_{a,sym}$ of Eq. (19) are evaluated. Symmetric modes are assumed to interact with the rigid-body aerodynamics by the longitudinal terms within Eq. (17), $C_{X\eta_k}$, $C_{Z\eta_k}$, $C_{m\eta_k}$ and also $C_{X\dot{\eta}_k}$, $C_{Z\dot{\eta}_k}$, $C_{m\dot{\eta}_k}$.

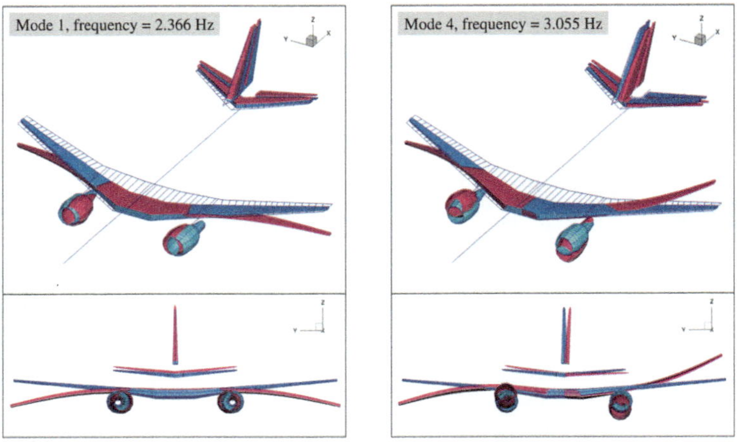

(a) First symmetric wing bending mode (b) First asymmetric wing bending mode

Fig. 7 Example of two different aircraft wing bending modes of the Airbus A320 configuration

The first wing bending mode of the Airbus A320 configuration is shown as an example of a symmetric mode in Fig. 7a.

Lateral-Directional Mode Shape (Asymmetric) Lateral-directional modes consist of the asymmetric wing bending modes but also any other engine or empennage mode that is not symmetric to the x-z-plane. Figure 7b shows the example of an asymmetric wing bending mode of the Airbus A320.

Using this categorization, only the lateral-directional components of Eq. (19) are supposed to be necessary to model the structural dynamics, which are β, p^*, r^*, δ_a, δ_r. It is assumed that the lateral-directional modes do not interact with the symmetric modes and that the asymmetric modes only interact with the lateral-directional rigid-body aerodynamics via the terms $C_{Y\eta_k}$, $C_{l\eta_k}$, $C_{n\eta_k}$ and also $C_{Y\dot{\eta}_k}$, $C_{l\dot{\eta}_k}$, $C_{n\dot{\eta}_k}$.

B. Rigid-Body Aerodynamics Identification

The results of Sect. 5.1 (i.e., the identified values of the parameters) are used as inputs of the automatic process. These values are used in the initial steps of the process and are refined later on.

C. Iterative Process for the Structural Mechanics Identification and Interaction with Complete Aircraft Model

The model parameters to be estimated for the structural mechanics are the ones from Eqs. (19) and (20). The process for the structural mechanics identification is proposed as an iterative process (see Fig. 6). In each iteration cycle, Step 1 up to Step 5, one additional structural mode is included in the identification process. In the following, a description of each step of the iteration process for the identification of the structural mechanics and its interaction with the complete aircraft model is shown.

Step 1: Starting values for structural mechanics In order to avoid unexpected interruptions in the automatic identification process, reasonable starting values for the parameters which will describe the structural mechanics must be provided. The definition of these starting values is based on the regression for the model parameters within $Q_k(t)$, Eqs. (19) and (20). The individual steps performed within Step 1 of the automatic process are described in the following.

Step 1.1: Calculation of $Q_k(t)$ The "measured" generalized load $Q_k(t)$ acting on the k-th structural mode is not directly contained in the VFT data. However, it can be calculated from the last equation of Eqs. (3), assuming an undamped structural system (i.e. $\zeta_k = 0$), using the generalized mass m_k, the in-vacuo natural frequency ω_k and the "measured" time signals of the generalized displacement η_k and its second derivative $\ddot{\eta}_k$. The calculated $Q_k(t)$ is then used in Step 1.2.

Step 1.2: Regression for model parameters within $Q_k(t)$ With $Q_k(t)$ being calculated via Step 1.1, a regression can be applied to solve the equations for the model parameters, i.e. Eqs. (19) and (20). In Eq. (20) Step 1.2 accounts only for the influence of the current mode, e.g., by the parameters $C_{\eta_2\eta_2}$ and $C_{\eta_2\dot{\eta}_2}$, and the influence of the lower-order modes, e.g., by the parameters $C_{\eta_2\eta_1}$ and $C_{\eta_2\dot{\eta}_1}$, on the generalized load $Q_k(t)$. The regression does not account for higher-order structural modes since they are only included during further iterations.

For the regression, the input data (α, β, p, q, r, $\eta_1 ... \eta_k$, $\dot{\eta}_1 ... \dot{\eta}_k$ and all control surface deflections) are taken directly from the CFD/CSM simulations.

Step 1.3: Contribution of model parameters to $Q_k(t)$ The linear regression provides model parameter values for all currently included parameters. However, it is to be expected that some regressors have only a negligibly small or no effect on $Q_k(t)$. The inclusion of such parameters which describe only a very small influence of a specific regressor on the overall generalized load of each mode k adds unnecessary complexity to the system and should be avoided. Therefore, the contribution of each selected parameter to $Q_k(t)$ is calculated in this step.

Each partial contribution of a selected parameter to the global generalized load $Q_k(t)$ is defined as $Q_{part}(t)$. For instance, the contribution of the parameter $C_{\eta_k\alpha}$ to the generalized load of the k-th structural mode is calculated by:

$$Q_{part,C_{\eta_k\alpha}}(t) = \bar{q}S\bar{c}C_{\eta_k\alpha}\alpha(t) \tag{21}$$

Step 1.4: Definition of main contributors to $Q_k(t)$ Step 1.4 contains the definition of the threshold for the contribution of each parameter to the generalized load equations, see Eq. (22).

The contribution $Q_{part}(t)$ is a time signal and its root mean square (rms) value relative to the root mean square value of the total signal is compared to a threshold which defines if a parameter is automatically included in the estimation or discarded and set to 0 in the current iteration step. The parameters $\boldsymbol{C}_{\eta_i\eta_i}$ and $\boldsymbol{C}_{\eta_i\dot{\eta}_i}$ are excluded from this step for all i and will never be discarded.

$$\frac{rms(Q_{part}(t))}{rms(Q_k(t))} > threshold \tag{22}$$

Parameters dropped from the equations using Step 1.4 can be included again later on in the process.

Step 2: Output error for structural mechanics of structural mode *k* Contrary to Step 1, where the generalized load $Q_k(t)$ was used as output to be matched and the parameters of a mode k were estimated by regression, from this step on the output error method (OEM) is used. The generalized displacements and their first and second derivatives are outputs to be matched. With this approach the fitting of the η_k and $\dot{\eta}_k$ can be improved. The parameter values of a mode k determined in Step 1.4 are used as initial values for the estimation. This includes all mode interaction parameters of lower-frequency modes on mode k deemed necessary in Step 1.4, but not the mode interaction parameters of mode k on the lower-frequency modes.

At the end of the estimation process for the structural mechanics with the cost function minimization completed, it is automatically decided if selected parameters are well identifiable or not. This means, that based on a threshold for the relative standard deviation of the estimated parameter (Cramér-Rao bounds) individual parameters may be dropped from the estimation process. Again, the parameters $\boldsymbol{C}_{\eta_i \eta_i}$ and $\boldsymbol{C}_{\eta_i \dot{\eta}_i}$ will never be dropped for all i. If parameters are dropped, Step 2 is restarted, up to a set of well identifiable parameters.

Step 3: Output error for mode interactions with lower-frequency modes The mode interaction describes the influence each structural mode might have on another mode. Since the process described in this work is iterative with increasing structural mode frequencies, this step adds to the selected parameters of mode k the interaction of the current mode k with the lower-frequency modes, e.g., $C_{\eta_1 \eta_2}$. The separation between longitudinal and lateral-directional modes is accounted for, which means that longitudinal modes only interact with other longitudinal modes. The same is valid for the lateral-directional modes, which only interact with other lateral-directional modes.

Whereas within Step 2 only the generalized displacement and its derivatives for structural mode k are used as outputs to be matched, in Step 3 the generalized displacements and the first- and second-order derivatives of all considered structural modes are used.

In the same way as in Step 2 it is automatically decided if individual parameters are dropped from the estimation process.

Step 4: Output error for complete structural mechanics The mode interaction parameters estimated in Step 3 may significantly affect the characteristics of a lower-frequency mode. Therefore, in this step, the parameter estimates obtained so far are used as starting values for the next computation, which is to perform the estimation of all accounted parameters of the structural mechanics. This means that contrary to Steps 2 and 3, where only the parameters of the mode k are accounted for, all retained parameters of the structural mechanics are estimated and updated. This assures the best identification of the complete structural mechanics with all identified modes.

In Step 4 the generalized displacements and the first- and second-order derivatives of all considered structural modes are used as outputs to be matched.

Again, in the same way as in Step 2 it is automatically decided if individual parameters are dropped from the estimation process.

Step 5: Output error for complete aircraft model The rigid-body dynamics model is represented by the parameters as estimated in Subsect. 5.1. The structural mechanics model is represented by all parameters identified in Step 4. The additional information in this step is the inclusion of the interaction between the structural dynamics of each structural mode and the rigid-body dynamics. Within this step all selected/desired parameters of the rigid-body dynamics (from Eqs. (10) to (15)) and the parameters which describe the interaction between structural dynamics and rigid-body dynamics (from Eq. (17)) must be reestimated. The rigid-body outputs (V_{TAS}, α, β, etc.) must be included as matching outputs in the estimation process. At the end of Step 5, it is assumed that an acceptable flexible aircraft model is identified including the structural dynamics. This automatically identified model is used as input in Step 1.1 if an additional structural mode is included in the model.

The above process is written in MATLAB and the OEM identifications are performed using a C++ Code for the aircraft model as described in Sect. 4. Regressors with an influence of less than 0.05% of $Q_k(t)$ were dropped and the threshold of the relative standard deviations of the parameters to be identified was set to 15%. Finally, OEM computation was stopped when the relative change of the cost function was less than 10^{-7}. Even though the flexible aircraft characteristics were included, they had only a negligible influence on the rigid-body dynamics and consequently the parameters which model these influences could not be identified. This was not unexpected, since the A320 is not considered a very flexible aircraft.

The complete identification process, for an aircraft model including 21 structural modes, consumed approximately 3 h of computation on a standard desktop PC (Intel(R) Core(TM) i7 CPU @ 3.40GHz, MATLAB 32bit 7.5.0 R2007b).

Exemplary results of the automatic process are shown in Fig. 8. The matches of the modal deflections of the symmetric mode 1 and the asymmetric modes 4 and 8 are very good for the VitAM and the VicToria maneuvers used for the identification. However, the match for the PoM maneuver proves that the quality of the identified model varies widely even in cases where the identification results indicate a very good model. It is still good for mode 4, for mode 1 it is probably sufficient for many applications, and for mode 8 it is not acceptable. Unfortunately, there are three modes where not even for the identification maneuvers a satisfactory match was obtained. They are shown in Fig. 9.

The reason(s) for the poor match could not be found within the time frame of the VitAM project. It could be a problem with the data, it could be a problem with the model, e.g. unmodeled nonlinearities due to transonic flow effects, there may be a better identification process, it could even be a programming error and there are several more potential reasons.

For an in-depth investigation the problem would need to be simplified and the information content of the identification maneuver should be improved in the frequency range of the elastic modes:

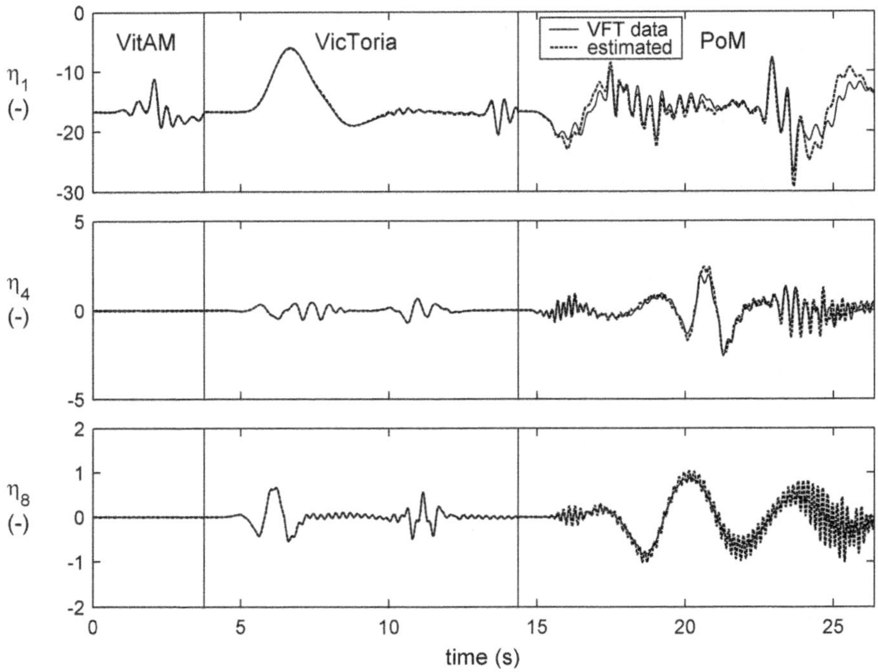

Fig. 8 Matches of modal deflections of modes 1, 4, and 8

- The number of elastic modes in the high-fidelity CFD/CSM model should be increased incrementally starting from one resulting in many additional high-fidelity simulations.
- The flight condition of the simulations should be less challenging, i.e. subsonic flow.
- A new maneuver with higher excitation frequencies should be designed.

6 Achievements and Outlook

A new method to design single- or multi-input signals that can be used as aircraft parameter estimation maneuvers was developed. Based on classical design criteria, the methodology allows the user to design complex signals both with and without a-priori information on the system to be identified (here the aircraft), even though only a quite restricted number of parameters is used to describe these signals. The proposed method was used to design two very short, multi-input maneuvers, which permitted to successfully estimate an aircraft's aerodynamic parameters from virtual flight test data.

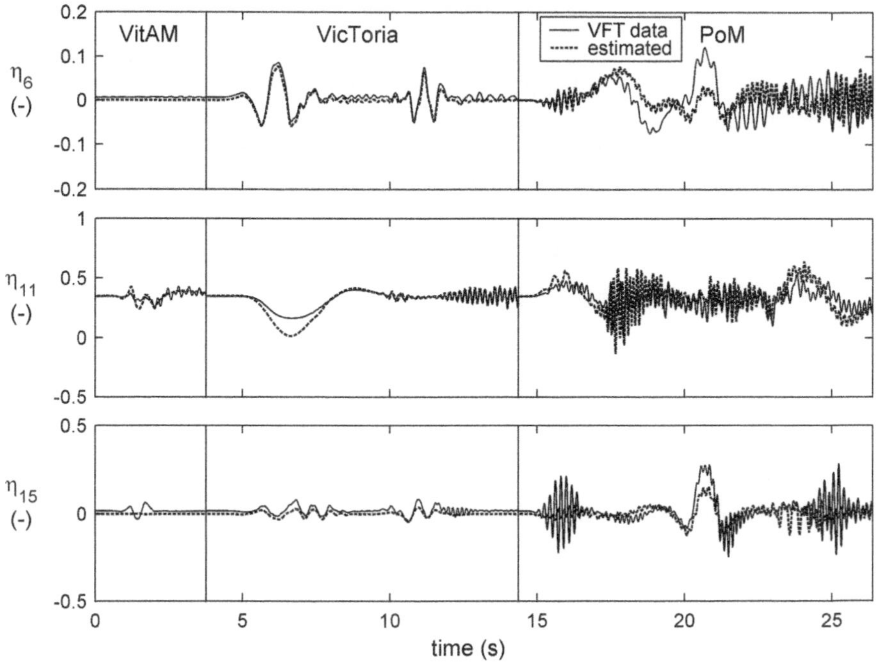

Fig. 9 Bad matches of modal deflections of modes 6, 11, and 15

An automatable process for the identification of a rigid-body aircraft model from virtual flight test data is proposed. It worked very well for the two example maneuvers, i.e. it resulted in a reasonable set of parameters and very good matches of all observation variables. It turned out that for the example aircraft it is not necessary to account explicitly for (dynamic) elastic effects on the rigid-body motion. Of course, effects of (quasi-) static elastic deformations are indirectly modeled by the aerodynamic parameters of the classic rigid-body model, whose values differ from those of a stiff model.

Validating the model by applying the identified model to a proof-of-match maneuver resulted in significantly worse matches of the observation variables. To investigate the reasons for the comparatively poor matches of the PoM maneuver would require additional high-fidelity simulations. Although the validation failed, it is remarkable that from such short maneuvers a rigid-body model for this trim point could be identified that matches the two identification maneuvers almost perfectly and a complex proof-of-match maneuver for large parts inside the generalized Level D tolerances.

An automatic process for the identification of a flexible aircraft model from virtual flight test data was developed and applied to two multi-axis system identification maneuvers simultaneously. This approach allows to identify models that can be used for real-time computations, control law design and handling qualities analysis, amongst others, with a much lower computation effort than a CFD/CSM solver.

In the investigated case, unfortunately, the process did not result in a fully useable model. Although the time histories of many of the 21 simulated modes were matched very well, the validation failed for three high(er) order modes. Depending on the application it may, however, be possible to reduce the model to the most important low-frequency modes to arrive at a suitable model. Investigating the reasons for the deficiencies of the identified model would require a considerable budget for many additional CFD/CSM simulations and the subsequent analysis and a serious commitment of all involved parties.

References

1. Flight Dynamics – Concepts, Quantities and Symbols – Part 1: Aircraft Motion Relative to the Air. Standard, International Organization for Standardization (ISO), Geneva, Switzerland (1988). ISO 1151-1:1988
2. Airplane Simulator Qualification. Standard, Federal Aviation Administration (FAA), Washington, DC, USA (1991). AC 120-40B
3. Jategaonkar, R.V.: Flight Vehicle System Identification - A Time Domain Methodology, 2nd edn. Progress in Astronautics and Aeronautics, vol. 245. AIAA, Reston, VA, USA (2015). ISBN: 978-1-62410-278-3
4. Jauberthie, C., Chanthery, E.: Optimal input design for a nonlinear dynamical uncertain aerospace system. In: Proceedings of the 9th IFAC Symposium on Nonlinear Control Systems, vol. 46, pp. 469–474 (2013)
5. Lichota, P.: Inclusion of the D-optimality in multisine manoeuvre design for aircraft parameter estimation. J. Theor. Appl. Mech. **54**(1), 87–98 (2016)
6. Lichota, P., Ohme, P.: Design and Analysis of new Multi Axis Input Manoeuvres for Aircraft Sys-ID. Technical Report DLR-IB 111-2014/46, DLR Institut für Flugsystemtechnik (2014)
7. Mönnich, W.: Ein 2-Punkt-Aerodynamikmodell für die Identifizierung. In: Symposium "Systemidentifizierung in der Fahrzeugdynamik". DFVLR Mitteilung 87-22, Paper No. 3.1 (1987)
8. Morelli, E.A.: Flight test of optimal inputs and comparison with conventional inputs. J. Aircr. **36**(2), 389–397 (1999)
9. Morelli, E.A.: Multiple input design for real-time parameter estimation in the frequency domain. In: IFAC Proc. Vol. **36**(16), 639–644 (2003). Proceedings of the 13th IFAC Symposium on System Identification (SYSID 2003)
10. Roeser, M.S., Fezans, N.: Method for designing multi-input system identification signals using a compact time-frequency representation. CEAS Aeronaut. J. **12**(2), 291–306 (2021)
11. Roeser, M.S., Mönnich, W., Ritter, M.: A process for time domain identification of a flexible aircraft model from CFD/CSM simulations. In: AIAA SciTech Forum (2022)
12. Waszak, M.R., Schmidt, D.K.: Flight dynamics of aeroelastic vehicles. J. Aircr. **25**(6), 563–571 (1988)

Numerical Studies on the Impact of Atmospheric Turbulence on Aircraft Loads

Jens Müller and Thorsten Lutz

Abstract A CFD simulation of the Airbus A320 ATRA encountering broadband atmospheric turbulence based on an isotropic von Kármán spectrum is presented and analyzed. The Disturbance Velocity Approach within the DLR flow solver TAU is used to represent the velocities of the atmospheric turbulence in the simulation. Statistical analysis of global and spanwise lift coefficients as well as local c_p spectra demonstrate the influence of atmospheric turbulence on the aerodynamic characteristics of the aircraft. The impact of the load fluctuations at the engine and the flap track fairings on the global and local lift spectra of wing and aircraft are evaluated. Pressure spectra on the wing's upper surface, evaluated at selected spanwise positions, allow a detailed assessment of the influence of flap track fairing and engine. Finally, the results are compared to previous investigations of a clean wing in atmospheric turbulence. It can be shown, that the aerodynamic response of the aircraft, relevant in terms of flight dynamics and passenger comfort, is well represented by the wing without considering the loads occurring on flap track fairings or engine. Since especially the engine has relevant contributions to the load spectra at higher wave numbers, it should be considered when evaluating the effect on the aircraft structure.

Keywords Aircraft aerodynamics · Atmospheric turbulence · CFD · Disturbance velocity approach

1 Introduction

The interaction of aircraft with atmospheric disturbances is relevant throughout the entire period of a flight mission. According to the EASA certification specifications (CS-25) both, discrete gust and continuous turbulence loads have to be considered within the aircraft design process [1]. For continuous turbulence loads, the CS-25

J. Müller (✉) · T. Lutz
Faculty 6: Aerospace Engineering and Geodesy, Institute of Aerodynamics and Gas Dynamics, University of Stuttgart, Wankelstraße 3, 70563 Stuttgart, Germany
e-mail: jens.mueller@iag.uni-stuttgart.de

R. Heinrich (ed.), *Advanced Aircraft Understanding via the Virtual Aircraft Model*,
Notes on Numerical Fluid Mechanics and Multidisciplinary Design 155,
https://doi.org/10.1007/978-3-031-69425-7_6

considers one dimensional vertical or lateral disturbances based on a von Kármán spectrum with turbulence length scale $L = 2500$ft. For different altitudes, the length scale is kept constant. Since this length scale is significantly larger than the wing's chord length, it mainly causes large scale variations in the inflow velocity vector. But, when flying at low altitudes and subsonic speeds, for example during takeoff and landing, the length scale of the atmospheric turbulence becomes smaller due to the atmospheric boundary layer. Thus, the effect of three-dimensional atmospheric turbulence, with length scales in the order of magnitude of the local chord length, on the wing's aerodynamic response becomes relevant. These small scale inflow velocity fluctuations lead to highly unsteady aerodynamic effects, affecting the aerodynamic performance as well as structural loads and passenger comfort. Most investigations considering the effect of small scale atmospheric turbulence on airfoils or wings, for example Hoffmann [6] or Hancock and Bradshaw [3], were carried out at low Reynolds numbers, which are not relevant for airliners. They also mainly focus on small length scales influencing the boundary layer. Other investigations like the one of Ohno et al. focus on the influence of continuous gusts on the boundary layer transition and laminar separation bubbles through a combination of URANS and direct numerical simulation [13]. A more detailed overview of recent research activities is provided in [12].

The objective of the present work is to evaluate the effect of three-dimensional broadband atmospheric turbulence, where the length scale of the turbulence spectrum is in the order of the mean aerodynamic chord of the wing, on the aerodynamics of an airliner, the Airbus A320 ATRA [14], by means of CFD simulation. The flight conditions are subsonic within the linear lift region of the aircraft. The effect of this kind of turbulence on the aerodynamics of an extruded airfoil [9] and a clean wing [12] of a short to medium range single aisle aircraft was investigated by the authors prior to the current investigations. As in the present work, CFD simulations were used as well. These investigations show, that leading edge and the suction peak of airfoil and wing are most affected by the atmospheric turbulence with a strong correlation between the c_p spectra at these positions and the inflow velocity spectrum of the atmospheric turbulence. The inboard region of the wing is less affected by the atmospheric turbulence than the outboard region due to the less pronounced suction peak at the inboard region. The correlation between the spectra of the atmospheric turbulence and the local pressure coefficient on the wing increases with increasing spanwise position. More details can be found in [9, 12]. The analyses presented in the work at hand continue this previous work. Instead of looking at an isolated airfoil or wing, the effects on a full aircraft configuration including wing, fuselage, engine, horizontal (HTP) and vertical tailplane (VTP) are now considered.

A key objective of the current work is to identify the contribution of different parts of the aircraft (FTF, engine, HTP) to the load fluctuations due to the interaction with atmospheric turbulence. By comparison of the A320 ATRA simulation with the results of the FNG wing in [12], the impact of engine and FTFs on the aerodynamic response to atmospheric turbulence is evaluated.

The numerical methods used for the simulation of the ATRA encountering atmospheric turbulence are described in Sect. 2. Details on the geometry itself, inflow conditions, and numerical parameters are presented in Sect. 3, while the results of the DVA simulation are discussed in Sect. 4. First, the global loads are evaluated in Sect. 4.1. Further insight into the contribution of the FTF and the engine to the load spectra are provided in Sect. 4.2 by an analysis of the spanwise lift coefficient. Finally, the evaluation of local pressure coefficients on the wing is presented in Sect. 4.3, which enables a further assessment of the differences between the inner wing, which is influenced by the engine and the FTFs, and the outboard wing.

2 Numerical Methods

Representing atmospheric turbulence involving a broad frequency range in CFD simulation is a key challenge. The most physical approach is to resolve and propagate the atmospheric turbulence within the flow field. Thus, the mutual interaction between the atmospheric turbulence and the aircraft can be covered. In CFD simulation this can be realized for example via modified far field boundary conditions [4] or a momentum source term [12], where the turbulence is fed into the flow field. The method is called Resolved Atmosphere Approach (RAA). Due to the broad frequency range of the atmospheric turbulence, it requires a highly resolved CFD mesh in the turbulence propagation region to avoid numerical dissipation and dispersion, making the simulation very expensive in terms of computational costs.

Another method is the so-called Disturbance Velocity Approach (DVA) [4], which is used for the CFD simulation in the present study. Here, the turbulence is not resolved and propagated in the flow domain. Instead, the disturbance velocities $\mathbf{u}'$ of the atmospheric turbulence are added to the flux balance of each cell by superposition. Hence, the disturbance velocities are treated similar to grid velocities. As an example, the continuity equations can now be written as

$$\frac{d}{dt}\int_V \rho\, dV + \oint_S \rho(\mathbf{u} - \mathbf{u}_b - \mathbf{u}') \cdot \mathbf{n}\, dS = 0 \tag{1}$$

with $\mathbf{u}_b$ being the velocity of the boundary of a control volume [4]. This changes the convection across the cell interface of a control volume from $\mathbf{u} - \mathbf{u}_b$ to $\mathbf{u} - \mathbf{u}_b - \mathbf{u}'$. The DVA enables the use of non-refined standard CFD meshes, significantly reducing the computational costs. In contrast to the RAA, the DVA only covers the interaction of the atmospheric turbulence with the aircraft but not the changes in the atmospheric turbulence due to the interaction with the aircraft. More details on the DVA can be found in [4, 5].

The applicability of the DVA for CFD simulations of aircraft interacting with atmospheric turbulence was widely validated for different purposes against the aforementioned RAA, which covers all mutual interactions between atmospheric turbulence and aircraft. In [4, 5] Heinrich and Reimer assessed the accuracy of the DVA for a 2D wing-HTP configuration in sub- and transonic flow and for a generic aircraft at transonic flight in terms of lift response to vertical gusts. Müller et al. continued the analyses of the DVA by Heinrich and Reimer in [10, 11]. They investigated the accuracy and applicability of the DVA for the interaction of a single 2D airfoil with vertical gusts at sub- and transonic flow conditions. Within their work, they focus on lift-, drag and pitching moment coefficient while gust wavelength, airfoil shape and angle of attack are varied. The results show that the DVA provides accurate results for gust wavelengths $\lambda_{gust} \geq c$ where c is the chord length. Relevant deviations in the DVA results compared to the RAA occur for smaller wavelengths as well as high angles of attack.

Concerning the interaction of airfoil and wing with broadband three-dimensional atmospheric turbulence, the accuracy of the DVA compared to the RAA was investigated by Müller et al. in [9, 12]. The wavelength of the von Kármán spectrum used for the atmospheric turbulence is in the order of the chord length respectively the mean aerodynamic chord of the wing. The studies show that the DVA covers the main aerodynamic effects on the airfoil/wing when interacting with atmospheric turbulence. Global and local lift coefficient as well as global pitching moment coefficient of the DVA agree well with the RAA results over a broad range of wave numbers. Mean pressure distributions and standard deviations match as well. The DVA's c_p spectra at the leading edge and the suction peak are accurate over the entire wave number range. With increasing x/c the DVA shows slightly higher amplitudes in the c_p spectrum at high wave numbers than the RAA. It is referred to [9, 12] for further details. Overall, the DVA provides reliable results for the interaction of aircraft with three-dimensional atmospheric turbulence. Therefore, this approach is used for the CFD simulation of the A320 ATRA encountering atmospheric turbulence within the present work.

3 Geometry, Setup, and Inflow Conditions

Based on the aerodynamic analysis of airfoil and wing encountering broadband three-dimensional atmospheric turbulence presented in [9, 12], the present work evaluates the impact of atmospheric turbulence on an airliner. The aircraft under consideration is DLR's Airbus A320-232 D-ATRA [14], a single aisle commercial aircraft with lift improvement package [15]. The CFD simulation includes fuselage, wing, flap track fairings (FTF), engine, vertical tailplane (VTP), and horizontal tailplane (HTP) as shown in Fig. 1. The wing is considered in cruise configuration, with flaps and slats retracted.

The flight conditions are subsonic at $M = 0.289$ and $Re_{c_\mu} = 21.8 \cdot 10^6$. The reference chord c_μ of the wing is 4.19 m and the wing's reference area $S_{ref} = 122.4\text{m}^2$. Further details regarding the aircraft geometry are provided in [15]. In line with the previous studies involving the FNG wing in [12], the angle of attack for the current investigation is set to $\alpha = 4°$, which corresponds to the linear lift region. Choosing the same angle of attack as in [12] allows a good comparison with these results. The moderate α also prevents separations to occur during the interaction of the aircraft

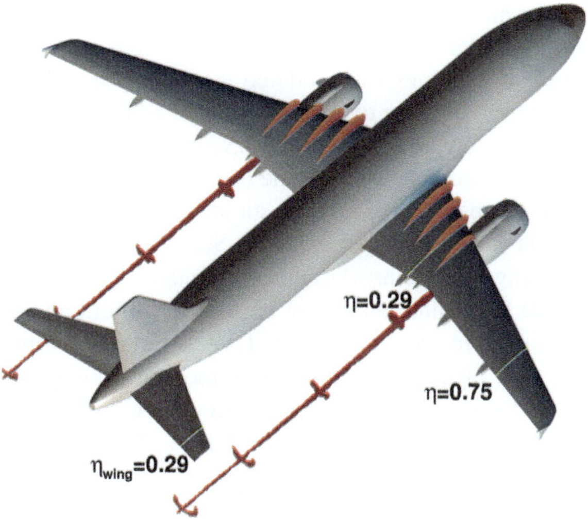

Fig. 1 Airbus A320 ATRA geometry with slice positions for detailed evaluations. Half model is used within the CFD simulations

with atmospheric turbulence. Investigation of the effects on separation bubbles is not the target of the current investigation.

The three-dimensional inflow turbulence, which interacts with the aircraft, is generated prior to the simulation based on the Mann model. The Mann model generates a three-dimensional velocity field, also called "Mann-box", which can be fed into the CFD simulation using the DVA. Details on the Mann model can be found in [7, 8]. As in [9, 12], the turbulence is based on the isotropic von Kármán spectrum. The length scale L of the von Kármán spectrum spectrum is 4.19 m, which corresponds to c_μ. The turbulence intensity of the atmospheric turbulence is set to 5% relative to the inflow velocity u_∞.

The compressible unstructured finite volume flow solver TAU [16], version 2018.1.0, developed by the German Aerospace Center (DLR) is utilized for the CFD simulations within this work. It provides second order accuracy in time and space. The DVA, which is used to represent the atmospheric turbulence field within the CFD simulation, was implemented into TAU by Heinrich and Reimer [4]. Within the simulation, a symmetric half model of the A320 ATRA is used. The CFD grid is a hybrid mesh consisting of approximately 46 million grid points. The grid was generated at the IAG within the LuFo project HINVA. Details on the grid generation can be found in [2]. The surface grids of fuselage, HTP, VTP, and the junctions of fuselage and wing as well as the ones of fuselage and HTP/VTP are fully structured in order to avoid collapses of the structured boundary layer cells in wall normal direction in the area of the junctions. Hence, the boundary layer consists only of hexahedrons in this area. Due to the increased complexity at the flap track fairings and the engine, the surface mesh of the outer wing, FTF, and engine consists of triangles which results in prisms in the boundary layer. $y^+ \leq 1$ is ensured within the boundary layer. Due to stability reasons when using the DVA with a three-dimensional atmospheric turbulence field, an Euler wall boundary condition is applied at the symmetry plane of the aircraft. Within the flow solver TAU the wall pressure correction is applied for the Euler wall as recommended when using Euler walls as symmetry plane. The thrust of the engine is modeled via the engine inflow and engine exhaust boundary condition. In order to evaluate the interaction of the atmospheric turbulence with the A320 ATRA, an unsteady RANS simulation was conducted. Within the simulation, a second order central scheme is used for the spatial discretization. The second order time stepping is realized by an implicit backward Euler scheme combined with a Lower-Upper Symmetric Gauss-Seidel scheme. The RANS equations are closed by utilizing the Spalart-Allmaras turbulence model in its original form [17]. Previous to the DVA simulation, the base flow without atmospheric turbulence present was simulated in order to start the DVA simulation with a fully converged base flow. At the start of the DVA simulation, the atmospheric turbulence is initialized with the most downstream front of the atmospheric turbulence field being located one length scale L upstream from the aircraft nose. Within the DVA simulation, the atmospheric turbulence propagates with the freestream velocity u_∞. In order to ensure an accurate temporal resolution of the atmospheric turbulence interaction, the simulation time step is set to $\Delta t = \frac{L}{u_\infty 150}$, similar to the simulations in [9, 12]. 150 inner iterations are performed within one time step. The statistical evaluations of the aircraft aerodynamics, presented in the following chapters, include 5400 time steps, which is equivalent to 36L passing over the wing. The statistical evaluations start when eight length scales have already interacted with the aircraft, in order to avoid the initial interaction of aircraft and turbulence to be taken into account.

4 Influence of Atmospheric Turbulence on the A320 ATRA

As described in Sect. 3, the aerodynamic response of the A320 ATRA to broadband isotropic atmospheric turbulence is analyzed in the following. The length scale of the von Kármán energy spectrum, used for the generation of the atmospheric turbulence, is equal to the reference chord length of the aircraft. In line with the findings in [12], variations in the inflow velocity vector are expected to be the main impact on the aircraft. First, the effects of the atmospheric turbulence on the aircraft's global aerodynamic loads are analyzed in Sect. 4.1. Subsequently, the spanwise variations of the local lift are evaluated in Sect. 4.2. The analyses are completed in Sect. 4.3 with an evaluation of the local c_p fluctuations on wing and HTP. Based on one CFD simulation of the A320 ATRA including all attached parts (engine, FTF, HTP, VTP etc.), the contribution of different components to the aerodynamic coefficients and their statistics is examined as well. Since no simulation without the attachments was performed, a separated comparison of the flow characteristics with or without attachments like engine end pylon is not part of the current work.

4.1 Global Aerodynamic Coefficients

The time history of lift and pitching moment coefficient of the A320 ATRA interacting with the atmospheric turbulence is shown in Fig. 2. c_μ is the reference length for the pitching moment coefficient. The values in Fig. 2 are normalized with the lift and pitching moment coefficients without atmospheric turbulence present from a previous steady RANS CFD simulation. Besides the lift coefficient $c_{L,\text{aircraft}}$ of the entire aircraft, separate c_L values of wing with FTF, wing with FTF & engine, and HTP are displayed in the figure. For $c_{L,\text{wing\&FTF\&engine}}$ only the outer part of the engine (pylon and nacelle) is taken into account. The region, where the flow goes through the inner part of the engine (inflow, outflow region, spinner, plug etc.), is not considered for this specific c_L value. All values are extracted from the same CFD simulation of the ATRA interacting with atmospheric turbulence, where the entire aircraft geometry was considered.

When comparing the lift fluctuations due to atmospheric turbulence of the wing with and without the engine considered in the c_L, it becomes clear, that engine and pylon do not lead to a relevant increase in the normalized lift fluctuations. The characteristic of the normalized c_L history remains almost unchanged, with only small local deviations. Thus, regarding lift, the engine is not more prone to atmospheric turbulence than the wing. Since the steady reference values of the wing with and without engine do not differ significantly as well, the influence of engine and pylon on lift and lift fluctuations of the aircraft are negligible. The normalized lift of the entire aircraft exhibits locally slightly larger fluctuations than the wing. Overall, the normalized lift history agrees well with that of the wing in terms of characteristics and magnitude. In contrast to the wing and the entire aircraft, the relative lift signal of the HTP differs. In the steady reference case, without atmospheric turbulence

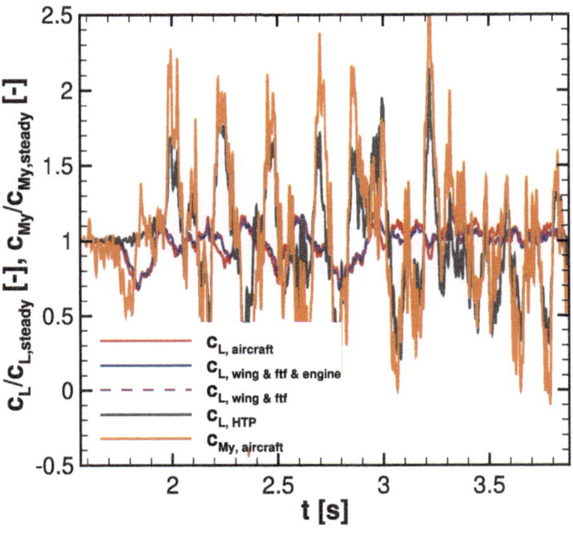

Fig. 2 Time histories of lift and pitching moment coefficients resulting from the interaction of the A320 ATRA with atmospheric turbulence. The names correspond to the components considered within the calculation of the particular coefficient

present, the HTP produces a small downforce, which results in a nose-up pitching moment. The small magnitude of the HTP's downforce at steady state results in the relative fluctuations due to atmospheric turbulence being significantly larger than is the case for the wing or the entire aircraft. Since the HTP exerts a high influence on the aircraft's pitching moment around the center of gravity, due to the long lever arm, the characteristic as well as the relative fluctuations of the pitching moment are similar to the ones of the HTP's lift.

The spectral components of the different lift coefficients from Fig. 2, as well as the pitching moment coefficient spectra are shown in Fig. 3 by utilizing a Fast Fourier Transform (FFT). As mentioned in Sect. 3, the initial interaction of the aircraft with the atmospheric turbulence is not considered within the statistical evaluation. The amplitudes are plotted against the normalized wave number $\kappa = \frac{2\pi f L}{u_\infty}$.

All spectra show locally higher amplitudes for $2 \leq \kappa \leq 3$. A second increase in amplitudes is visible between $6 \leq \kappa \leq 23$, which is visible as a plateau in the corresponding wave number range. The plateau extends from $7 \leq \kappa \leq 23$ in case of the lift spectra of the entire aircraft (Fig. 3a). When considering only the wing with FTF in the lift coefficient, the plateau ranges from $\kappa = 9$ to $\kappa = 18$, whereas it ranges from $\kappa = 9$ to $\kappa = 23$ before the amplitudes decrease again, when wing, FTF, pylon and nacelle (engine) are considered in the c_L. This indicates an influence of the flow around the engine on the spectra at this wave number range. This aspect is discussed in more detail in Sect. 4.2.

Looking at the lift spectrum of the HTP and the pitching moment spectrum of the aircraft shown in Fig. 3b, the amplitude plateau mentioned above is visible between $6 \leq \kappa \leq 21$ for the lift coefficient of the HTP and between $7 \leq \kappa \leq 20$ for the aircraft's pitching moment coefficient. Since the aircraft's pitching moment is directly coupled with the lift fluctuations at the HTP, this similarity is well explained.

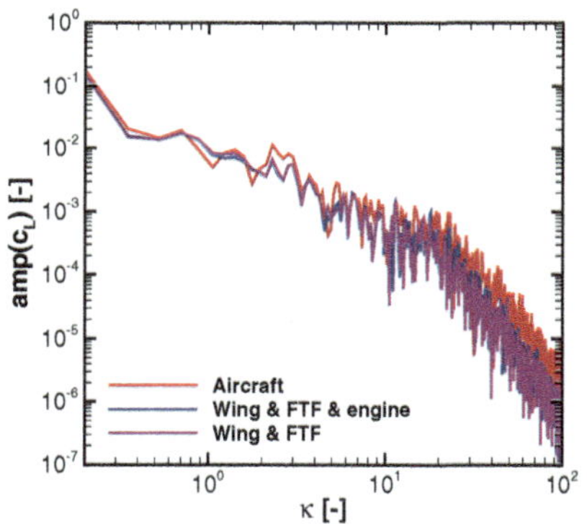

(a) c_L spectra of aircraft, wing with engine, wing without engine

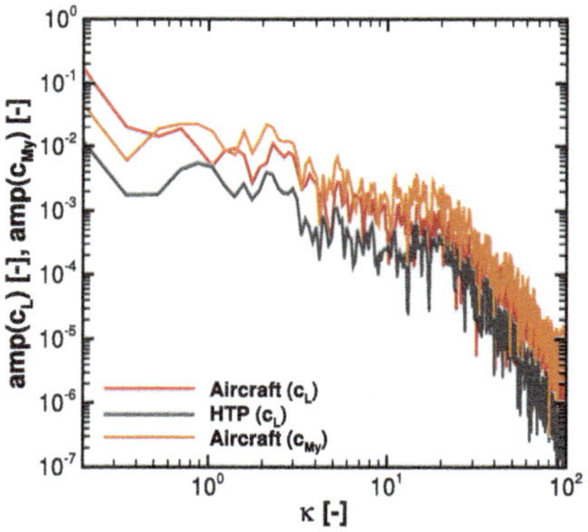

(b) c_L spectra of aircraft and HTP, c_{M_y} spectrum of aircraft

Fig. 3 Spectra of the global aerodynamic coefficients obtained by FFT

The slightly different wave number ranges of the plateau in the spectra of the different components indicate, that for the lower wave numbers ($6 \leq \kappa \leq 9$), the HTP as well as the pitching moment have a visible influence on the aircraft's lift spectrum. From $\kappa = 9$ to $\kappa = 18$, where all curves show a plateau in the amplitudes, it is not possible to determine a clear dominant influence of a specific component. Hence, all components (Wing, FTF, engine, HTP) contribute within this wave number range, since they all show a similar behavior in the spectrum. At higher wave numbers ($21 \leq \kappa \leq 23$), there is a visible influence of the engine on the lift spectrum of the aircraft.

Besides the two wave number ranges addressed above, the lift spectra of aircraft and wing with FTF without engine differ marginally. This was already evident in the time history of the lift coefficient. The lift spectrum of the aircraft has slightly higher amplitudes at higher wave numbers, which is consistent with the overall higher lift of the aircraft compared to only considering the wing with FTF for c_L. There is barely no difference over the entire wave number range in the amplitudes of the wing with and without pylon and nacelle of the engine, except between $18 \leq \kappa \leq 23$ as mentioned in the above. This shows, that within the simulation of the entire aircraft in atmospheric turbulence, the engine contributes only an insignificant part to the lift spectrum of the wing over a broad range of wave numbers.

The characteristics of the aircraft's pitching moment spectrum as well as the lift spectrum of the HTP, shown in Fig. 3b, are similar to the lift spectrum of the aircraft as well. The amplitude offset, especially between the lift spectrum of HTP and aircraft is associated with the lower lift of the HTP and the associated lower energy of the HTP lift signal.

It could be shown, that for the lower frequencies, relevant for flight dynamics and passenger comfort, the impact of the load fluctuations on the flap track fairings and the engine on the global loads of the aircraft is negligible. The large scale fluctuations in the lift's time history of the aircraft are mainly induced by the wing geometry, while FTF and engine have only minor contributions to these large scales. Regarding the aircraft's pitching moment, the main impact due to atmospheric turbulence is related to the HTP. For higher frequencies, there is a noticeable influence by the lift fluctuations on the engine geometry to the global lift coefficient of the wing including engine and FTF in the c_L calculation. Thus, the loads occurring on the engine geometry become relevant when higher frequencies have an impact on the aircraft structure but can be neglected in terms of flight dynamics.

4.2 Spanwise Lift Distributions

Having evaluated the global lift and pitching moment spectrum of the aircraft, together with a detailed look on the contribution of the wing as well as the engine on the lift response of the aircraft, the influence of the atmospheric turbulence on the lift coefficient of the wing is assessed in more detail in this section. Besides the general influence of the atmospheric turbulence on the lift coefficient of the wing, the

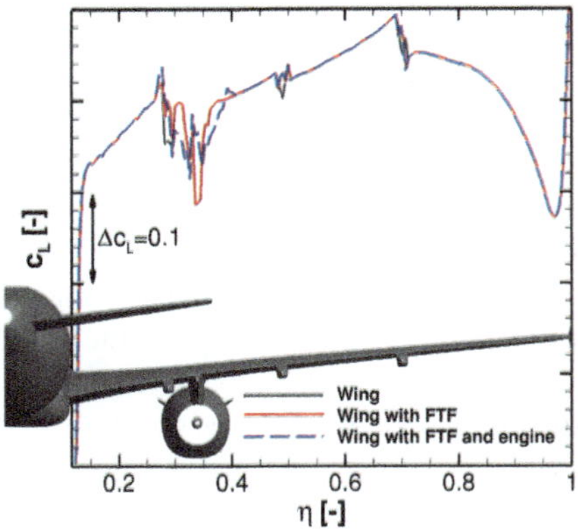

(a) mean value

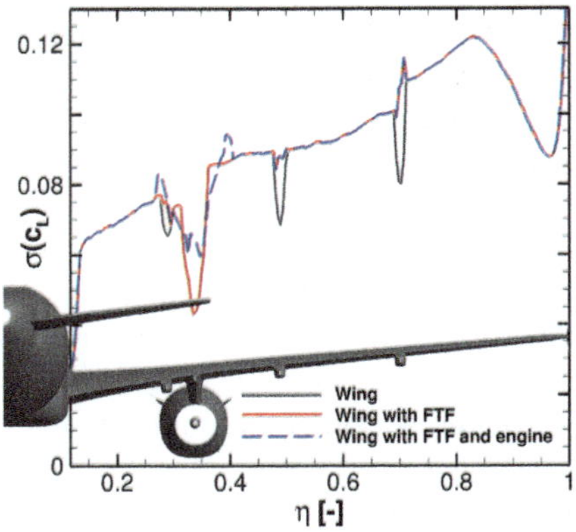

(b) standard deviation

Fig. 4 Time averaged lift coefficient with corresponding standard deviations of wing, wing with FTF, and wing with FTF and engine considered in the c_L calculation

contributions of flap track fairing and engine to the lift coefficient are evaluated. For this purpose, the spanwise lift coefficient of the wing is analyzed. Figure 4 shows the time averaged spanwise lift coefficient of the wing together with the corresponding standard deviations of c_L over the spanwise coordinate $\eta = \frac{y}{0.5b}$. b is the wing span of the aircraft. The standard deviations $\sigma(c_L)$ represent the local lift fluctuations due to the interaction with the atmospheric turbulence. The c_L values of only the wing, the wing with flap track fairings, and the wing with flap track fairings, pylon and nacelle (engine) are shown in the figure in order to evaluate the contribution of the different components to the lift coefficient of the entire wing within the DVA simulation of the A320 ATRA. The reference length for the spanwise lift coefficient is always the local chord length at the corresponding spanwise position of the wing without flap track fairings and without engine. This enables a good comparison between the different elements considered in the respective c_L values. The local chord length instead of the aircraft's reference chord is chosen in order to evaluate the influence of the atmospheric turbulence on the local airfoil aerodynamics.

From the engine to the outer flap track fairing, c_L as well as $\sigma(c_L)$ increase with increasing spanwise position η. The more outboard on the wing, the more pronounced the suction peak in the local pressure distribution. Since higher suction peaks are more prone to the inflow velocity fluctuations induced by the atmospheric turbulence, the standard deviations increase in line with c_L. This corresponds to the results of the clean FNG wing presented in [12]. Inboard from the engine, the local airfoil shape differs significantly in order to compensate the crossflow induced by the engine and the interference between wing and fuselage. This results in lower $\sigma(c_p)$ at the suction peak inboard from the engine, despite the suction peak being more pronounced in this area. In addition, the region of very low pressure at the suction peak extends over a smaller percentage of the local chord. Therefore, $\sigma(c_p)$ decreases faster with increasing x/c than outboard from the engine. This, in combination with lower pressures on the pressure side of the wing in the inner region of the wing, compared to the region outside the engine, results in the lower c_L and $\sigma(c_L)$ values between fuselage and engine. Outside from $\eta = 0.825$, the significant drop in $\sigma(c_L)$ agrees well with the likewise substantial drop in the spanwise lift coefficient. Hence, this is mainly related to the reduced suction peak with increasing η in this area, which is less prone to the inflow velocity fluctuations due to the atmospheric turbulence reducing $\sigma(c_L)$. Between $\eta = 0.7$ and $\eta = 0.825$ $\sigma(c_L)$ still increases, while the mean c_L decreases. One aspect leading to this phenomenon is the change of the flow at the most outboard flap track fairing. In order to illustrate this aspect, the standard deviation of the surface pressure coefficient c_p on the pressure side of the wing in the outer region of the wing is shown in Fig. 5a together with the time averaged streamlines on the surface. Inboard from the FTF, the regions of high $\sigma(c_p)$ are shifted to higher x/c with increasing η. The interdependence of the FTF with the flow around the wing leads to a local shift of the high $\sigma(c_p)$ values outboard from the FTF to more upstream positions relative to the local chord length. When integrated along the chord line, this increases $\sigma(c_L)$ while the mean c_L values decrease slightly.

The contribution of the flap track fairing itself to the lift coefficient in Fig. 4 is evaluated by comparing the values, where only the wing geometry is included in the calculation of the lift coefficient, to the values where wing and flap track fairing geometry are included in the lift coefficient calculation. The time averaged spanwise c_L values depicted in Fig. 4a indicate only a minor influence of the flap track fairings to the local lift. In contrast, the standard deviations in Fig. 4b reveal, that the flap track fairings have a significant contribution to the local lift fluctuation due to the interaction with atmospheric turbulence, especially at the more outboard region of the wing. This is also visible in Fig. 5a. There are high $\sigma(c_p)$ values at the junction of FTF and wing. As indicated by the streamlines, which drop from the

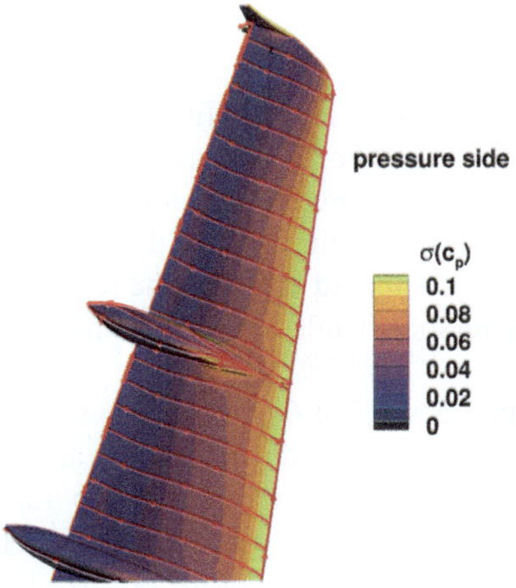

(a) outboard region (pressure side)

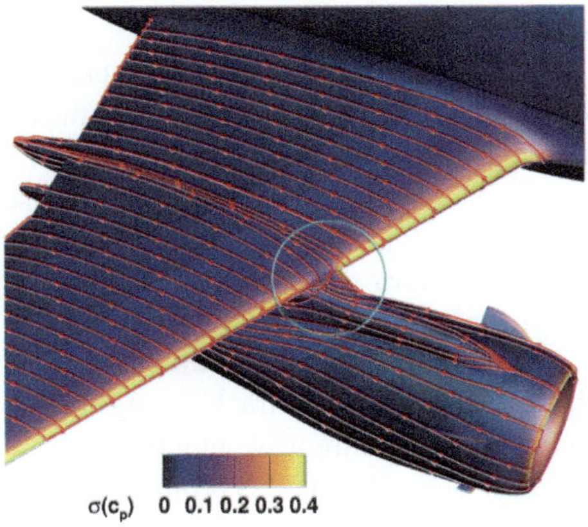

(b) inboard region

Fig. 5 Standard deviation $\sigma(c_p)$ with streamtraces based on the mean friction coefficient center of the FTF towards the wing with increasing x/c, the flow is accelerated on the outer sides of the FTF. This effect is more significant on the outboard side of the FTF due to outboard directed velocity component induced by the wing sweep. Since this occurs on the pressure side, the mean c_L value shows a significant local decrease at this spanwise position. Because low pressure regions are more prone to the inflow velocity fluctuations induced by the atmospheric turbulence, $\sigma(c_p)$ is locally increased as visible in Fig. 5a. These higher $\sigma(c_p)$ on the outboard surface of the FTF also explain the local peak in $\sigma(c_L)$ visible at the position of the most outboard flap track fairing in Fig. 4b.

When the engine is included in the c_L calculation, the changes in the mean c_L value and the standard deviations are significant. The presence of the pylon and engine change the local inflow conditions for the wing in this region. Hence, the suction peak near the leading edge of the wing is significantly decreased. This explains the general decrease in lift produced by the wing (or wing with FTF) at the spanwise position of the engine. Figure 5b shows that the decreased suction peak corresponds with significantly lower $\sigma(c_p)$ at the leading edge of the wing as highlighted by the cyan circle. As also visible in Fig. 5b, the acceleration of the flow near the stagnation points on the engine geometry (for example near the engine lip and the strakes), which corresponds with a higher sensitivity to inflow velocity fluctuations due to atmospheric turbulence, leads to an increase in $\sigma(c_p)$. Within the spanwise $\sigma(c_L)$ shown in Fig. 4b, this is reflected by local peaks in $\sigma(c_L)$.

For further evaluation, a FFT is applied to the spanwise lift signal. Contour plots of the lift amplitudes over the spanwise position η and the wave number κ are presented in Fig. 6. Figure 6a shows the spanwise lift coefficient spectra when only the wing is considered in the lift coefficient calculation. The spanwise c_L spectra with wing and flap track fairing taken into account in the c_L calculation is depicted in Fig. 6b. Finally, Fig. 6c displays the spanwise c_L spectra when wing, flap track fairing and engine are included in the computation of the spanwise c_L. For further clarification, the components not included in the c_L calculation are displayed transparent in each subfigure.

The overall characteristic of the spanwise distribution of the local lift spectra, with wing, FTF, and engine considered in the determination of c_L is displayed in Fig. 6c. Especially at lower wave numbers, lower c_L amplitudes are present in the inboard region of the wing. They increase outboard from the engine, where the standard deviations $\sigma(c_L)$ increase as well. With increasing outboard position, the region of high amplitudes shifts slightly to higher wave numbers while lower c_L amplitudes occur for $\log_{10}(\kappa) < 0.5$ ($\kappa < 3.16$).

Comparing the spanwise lift coefficient amplitudes in Fig. 6a, b it becomes clear, that neglecting the flap track fairings in the c_L calculation results in lower c_L amplitudes over the entire wave number range at the spanwise positions of the flap track fairings. This effect is less pronounced at the most inboard flap track fairing, which corresponds well to the reduced influence of the flap track fairing lift on $\sigma(c_L)$ at this position. The influence of the lift produced at the engine geometry can also be seen in the contour plot of the spanwise lift spectrum in Fig. 6c. As mentioned in Sect. 4.1, the different spectra of the integrated c_L, depending on the geometry

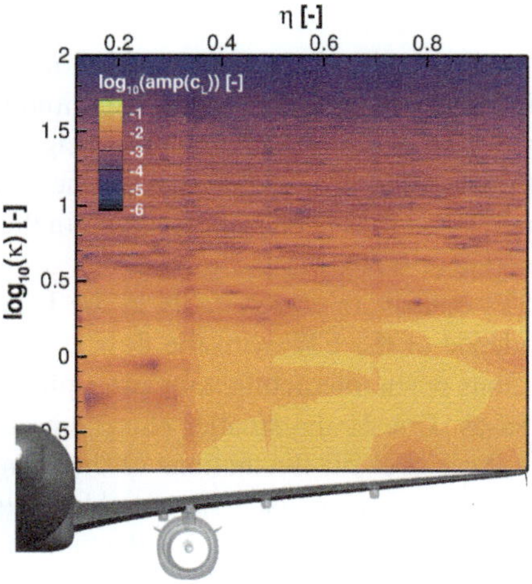

(a) wing without FTF and engine

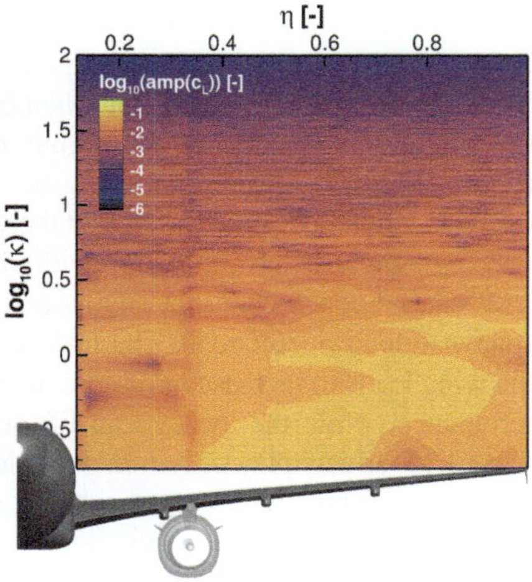

(b) wing with FTF and without engine

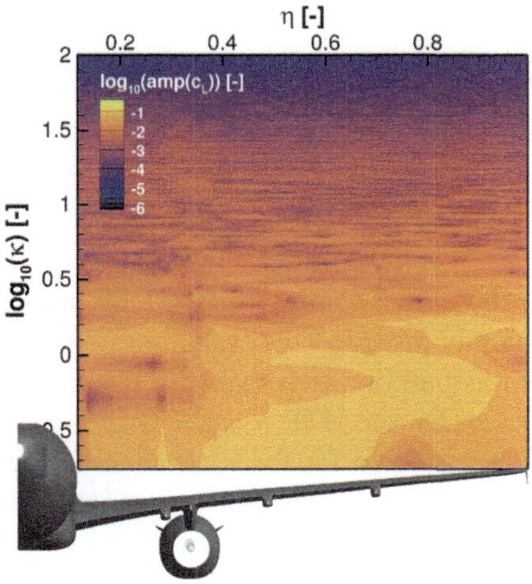

(c) wing with FTF and engine

Fig. 6 Spectral maps of the local c_L plotted over the spanwise position η. Geometries not included in the calculation of the spanwise c_L are displayed transparent

parts considered in the c_L determination, reveal an increase in c_L amplitudes for $6 \le \kappa \le 23$. For higher wave numbers, there is a clear decrease in the amplitudes. This decrease occurs at slightly lower wave numbers when the engine was not considered in the c_L value, which indicates an influence of the flow around the engine on this plateau. On detailed analysis of Fig. 6c, this influence can be confirmed. At the spanwise position of the engine, the local c_L amplitudes between $11 \le \kappa \le 28$ $(1.04 \le \log_{10}(\kappa) \le 1.45)$ are clearly higher than at lower/higher wave numbers. This amplitude increase in this wave number range is limited to the spanwise extent of the engine, as clearly visible in Fig. 6c. The fact that the influence originates from the engine can be confirmed by a comparison with Fig. 6b, where the geometry of the engine is not taken into account when calculating c_L. Here, the higher amplitudes at the spanwise position of the engine between $1.04 \leq \log_{10}(\kappa) \leq 1.45$ are significantly less pronounced than in Fig. 6c. Of course, the flow around the engine and the FTFs influences the flow around the wing, which cannot be separated within this investigation since no simulation without engine or FTFs was performed.

It can be concluded, that engine and FTFs contribute considerably to the local spanwise load fluctuations. The interdependence of the FTF with the flow around the wing leads to increased pressure fluctuations when interacting with atmospheric turbulence, while the local mean lift decreases. Due to the smaller chord length, these FTF effects play only a minor role in the global lift fluctuations of the wing or the aircraft but are worth considering when the impact on local structural loads is to be evaluated. A different situation occurs at the engine. The local lift fluctuations occurring on the engine/pylon geometry, especially at higher frequencies, have a clear influence on the spectra of local and global lift coefficient. Hence, this influence is worth to be considered for both, the global aerodynamic response of the aircraft as well as for local and global structural loads.

4.3 Local c_p Distributions

A deeper insight into the aerodynamic effects of the atmospheric turbulence on the wing's aerodynamics is provided by evaluating the local pressure coefficient on wing and HTP. For this purpose, two spanwise positions on the wing, $\eta = 0.29$ and $\eta = 0.75$ are selected. On the HTP $\eta_{wing} = 0.29$ is selected for a detailed c_p analysis. The positions considered for the c_p evaluation are also shown in Fig. 1. While $\eta = 0.29$ is located at the position of the most inboard flap track fairing, and hence covers the influence of flap track fairing and engine, the outboard position is comparable to a clean wing without the influence of flap track fairings or engine, as analyzed in [12].

The first assessment of the local influences of the atmospheric turbulence on the surface pressure coefficient c_p is made on the basis of the time averaged c_p values at $\eta = 0.29$ and $\eta = 0.75$. These c_p distributions of wing and HTP with corresponding standard deviations are shown in Fig. 7.

The standard deviations σ, displayed as bars in Fig. 7, show, that the pressure fluctuations on the wing due to the interaction with atmospheric turbulence increase with increasing spanwise position η. This corresponds well to the findings of Müller et al. in [12], where the impact of atmospheric turbulence on the clean FNG wing is analyzed. Thus, also in the case of the A320 ATRA with flap track fairings and engine, the outer region of the wing is most affected by the atmospheric turbulence as already pointed out in Sect. 4.2. Considerably smaller standard deviations of the surface pressure coefficient occur on the HTP than is the case on the wing. Both on the wing and on the HTP, the c_p fluctuations are highest at the leading edge and the

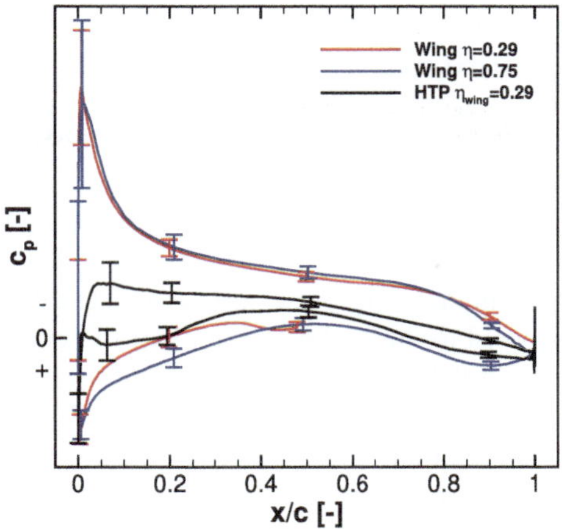

Fig. 7 Time averaged c_p and corresponding standard deviations at different spanwise positions on the wing and the HTP

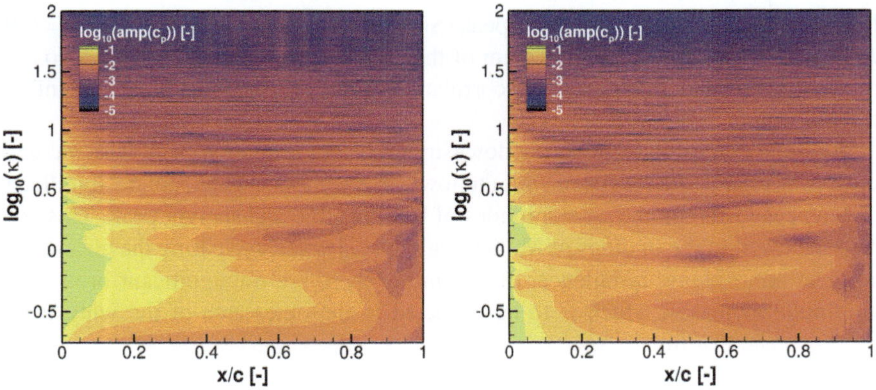

(a) $\eta = 0.75$ (b) $\eta = 0.29$

Fig. 8 Spectral maps of c_p on the upper surface of the wing plotted over the normalized chord position x/c

suction peak. With increasing downstream position x/c, the c_p fluctuations due to the interaction with atmospheric turbulence decrease significantly.

Applying a FFT on the c_p distributions at $\eta = 0.29$ and $\eta = 0.75$ enables a more detailed assessment of the influence of the atmospheric turbulence on the aerodynamic response of the A320 ATRA. Similar to Fig. 6, the c_p amplitudes along the chordwise lines on the upper surface of the selected spanwise cuts are plotted in Fig. 8.

First, the more outboard position on the wing at $\eta = 0.75$ (Fig. 8a) is analyzed. At this position, there is no interference with the fuselage, engine, FTF etc., so the c_p spectra should have characteristics comparable to the FNG wing results presented in [9, 12]. High fluctuation amplitudes can be observed near the leading edge at low wave numbers, which extend up to $\kappa \approx 5$ ($\log_{10}(\kappa) \approx 0.7$). At higher κ, the amplitudes decrease. With increasing downstream position x/c, the amplitudes decrease significantly, especially for low wave numbers. This corresponds very well to the findings of Müller et al. in [9, 12].

The influence of FTF and engine is assessed by comparing the c_p spectra on the wing at $\eta = 0.75$ with the ones at $\eta = 0.29$ shown in Fig. 8. At the inboard region, the overall lower c_p amplitudes (Fig. 8b) can be associated with the lower standard deviations $\sigma(c_p)$ compared to the outboard region of the wing, as shown in Fig. 7. There are local minima in the amplitudes present at $\kappa \approx 0.35$, $\kappa \approx 0.5$, $\kappa \approx 0.85$, $\kappa \approx 1.8$, and $\kappa \approx 2.5$ ($\log_{10}(\kappa) = \{-0,46, -0.30, -0.07, 0.26, 0.40\}$), which do not occur at the outboard region of the wing. This dedicates an influence of the flap track fairing and the engine on the c_p spectra on the suction side of the wing at this position. These peaks extend over almost the entire chord.

In order to assess the differences of the c_p spectra on the suction side of the wing in more detail, they are extracted at different chordwise positions for the two spanwise positions at the wing. Figure 9 shows the c_p spectra at $\eta = 0.75$ on the leading edge ($x/c = 0$), the suction peak (SP), $x/c = 0.2$, $x/c = 0.5$, and $x/c = 0.9$. Additionally, the u-velocity spectrum of the atmospheric turbulence (Mann-box) is plotted as well. This enables a good correlation of the local c_p spectra with the inflow turbulence.

Again, the amplitude reduction downstream from the suction peak with increasing x/c is clearly visible for medium to low wave numbers. There are no significant differences visible in the c_p amplitudes of leading edge and suction peak. This corresponds well to the findings of Müller et al. in [12] when evaluating the clean FNG wing in atmospheric turbulence. Furthermore, there are no significant broadband amplitude increases, which is to be expected in the outer area of the wing when

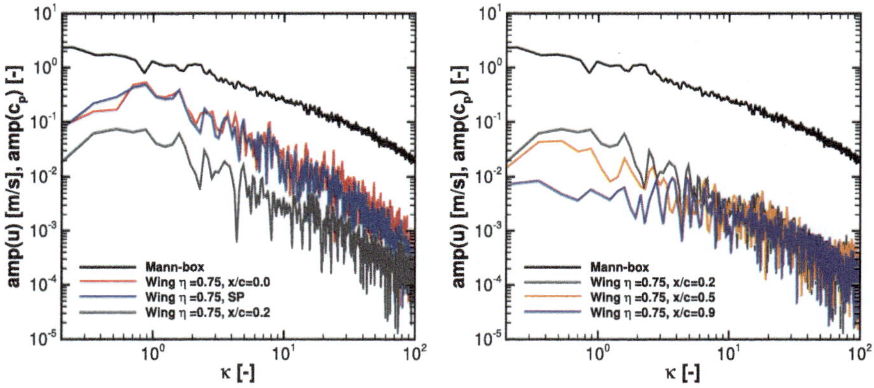

(a) $x/c = 0$, SP, $x/c = 0.2$ (b) $x/c = 0.2$, $x/c = 0.5$, $x/c = 0.9$

Fig. 9 c_p spectra on the upper (suction) side of the wing at $\eta = 0.75$

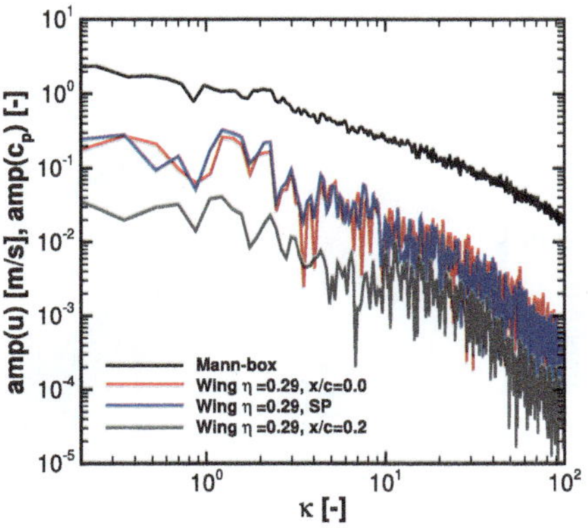

(a) $x/c = 0$, SP, $x/c = 0.2$

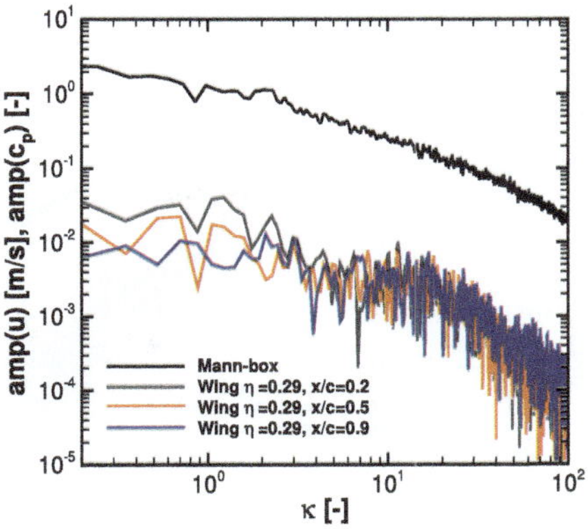

(b) $x/c = 0.2$, $x/c = 0.5$, $x/c = 0.9$

Fig. 10 c_p spectra on the upper (suction) side of the wing at $\eta = 0.29$

using the DVA as described in [12]. At all chordwise positions displayed in Fig. 9, amplitude peaks can be observed at $\kappa = 0.87$ and $\kappa = 1.57$. However, these peaks are not visible in the velocity spectrum of the atmospheric turbulence field (Mann-box). Despite these peaks, the characteristic of the chordwise c_p spectra correlate with the velocity spectrum of the atmospheric turbulence field. As stated in [12], this behavior is expected at the outboard region of the wing. Hence, there are no significant differences at $\eta = 0.75$ to the results of the clean FNG wing interacting with atmospheric turbulence presented in [12].

As already visible in the contour plot of the c_p spectra, the inboard cut at $\eta = 0.29$, which covers the influence of FTF and engine, shows a slightly different behavior than the outboard cut at $\eta = 0.75$. Figure 10 displays the c_p spectra at dedicated chordwise positions similar to $\eta = 0.75$ in Fig. 9. Between $\kappa = 3.67$ and $\kappa = 9$ as well as for $10 \leq \kappa \leq 21$, a plateau is present in the c_p amplitudes. In the aft region, at $x/c = 0.5$ and $x/c = 0.9$, the plateau in the c_p amplitudes extends over the entire range from $\kappa = 3.67$ to $\kappa = 21$. Especially the amplitude plateau at higher wave numbers ($10 \leq \kappa \leq 21$) corresponds well with the influence of the engine on the spanwise c_L distribution discussed in Sect. 4.2. A slightly different behavior is visible at $x/c = 0.2$. Here, an increase of the amplitudes for $7 \leq \kappa \leq 27$ is present in the c_p spectrum. Compared to the outboard cut at $\eta = 0.75$ and the investigation of the swept FNG wing in [12], this indicates a clear influence of the engine, the pylon and the flap track fairing on the c_p spectra on the upper surface of the wing. As already described, this is also reflected in the global and spanwise c_L spectra of the wing and the aircraft.

Finally, the local c_p spectra on the horizontal tailplane are evaluated in Fig. 11 for different chordwise positions. The spanwise position at the HTP, $\eta_{wing} = 0.29$ m corresponds to the one of the wing shown in Fig. 10. The chordwise positions on the HTP are located on the upper side. Since the HTP produces a downforce, only the c_p spectrum of the suction peak (SP) is located on the lower side of the HTP.

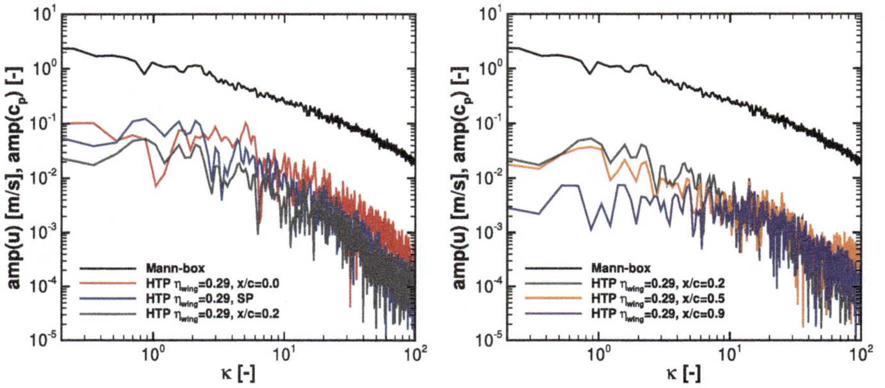

(a) $x/c = 0$, SP, $x/c = 0.2$ (b) $x/c = 0.2$, $x/c = 0.5$, $x/c = 0.9$

Fig. 11 c_p spectra on the upper (pressure) side and the suction peak (lower side) of the HTP at $\eta_{wing} = 0.29$

At the HTP's leading edge, a region of nearly constant c_p amplitudes is present for $1.6 \leq \kappa \leq 9$. The shape of the c_p spectrum corresponds to that of the inflow turbulence for $\kappa > 9$. Compared to the leading edge or the atmospheric turbulence, a steeper decrease of the c_p spectrum is visible at the suction peak. It starts at $\kappa = 4.4$ and continues for all higher wave numbers. This effect is considerably different from the c_p spectra on the wing of the A320 ATRA or the results of the FNG wing in [12]. On the other hand, the c_p plateau between $\kappa = 10$ and $\kappa = 21$ is also present on the upper side (pressure side) of the HTP, which is in line with the results on the wing. With increasing x/c, this region of nearly constant c_p amplitudes extends to higher wave numbers (up to $\kappa = 26$). The most downstream position on the HTP ($x/c = 0.9$) shows almost level c_p amplitudes also at low wave numbers. The decrease in amplitudes at small and medium wave numbers with increasing x/c, expected from the simulations of the FNG wing in [9, 12], and from the wing of the A320 ATRA remains evident on all chordwise positions evaluated on the HTP.

Overall, the evaluation of the local c_p fluctuations shows, that the impact of the atmospheric turbulence on the regions, where no influence of the FTFs on the flow is present, is comparable to the impact of atmospheric turbulence on a clean wing as studied in [12]. The presence of the engine and the FTF in the inboard region of the wing has a considerable influence on the local pressure fluctuations on the wings suction side with local minima in the c_p amplitudes at low wave numbers. At the outer part of the wing, the characteristic of the c_p spectra on the suction side correlates with the velocity spectrum of the atmospheric turbulence over a broad wave number range. This confirms that there are no relevant local aerodynamic effects affecting the interaction with atmospheric turbulence at this location. This is different at the inner region of the wing, where the flow on the suction side of the wing is influenced by the presence of the engine and the FTF. The differences in the load fluctuations on the engine for wave numbers $11 \leq \kappa \leq 28$ are also visible in the c_p spectra on the suction side of the wing. Hence, there is a strong correlation between the load fluctuations on the engine and the aerodynamic response on the suction side of the wing. This explains the only slight differences in the global lift spectra of the wing with and without engine considered in c_L presented in Sect. 4.1.

5 Conclusion

The impact of homogeneous atmospheric turbulence on the aerodynamics of a single aisle aircraft, the A320 ATRA, was investigated in subsonic flight. A CFD simulation utilizing the DLR TAU code was performed, where the velocities of the atmospheric inflow turbulence were represented using the Disturbance Velocity Approach. This approach covers the influence of the atmospheric turbulence on the aircraft but neglects any changes in the atmospheric turbulence due to the interaction with the aircraft. The DVA was already validated for this kind of simulation in [9, 12]. The work presented here builds on and extends the authors' previous work regarding the interaction of airfoils and wings with atmospheric turbulence within the LuFo project VitAM-Turbulence, published in [9–12].

A detailed analysis of the impact of the inflow turbulence on the A320 ATRA was performed. The contributions of the engine to the global lift spectra of the aircraft were identified. It could be shown that the impact of the load fluctuations on the flap track fairing and the engine is negligible when the focus is set on flight dynamics and passenger comfort. When load fluctuations at higher frequencies are relevant, for example regarding the impact on the aircraft structural dynamics, the geometry of the engine has to be included into the calculation of the aerodynamic loads. Spanwise evaluations of the wing's local lift coefficient revealed the influence of the pressure fluctuations on the flap track fairings and the engine on the local lift spectra. Engine and FTF contribute considerably to the local spanwise load fluctuations. Due to the smaller chord length, especially the influence of the FTF is not visible in the integrated global loads of wing and aircraft but relevant for the local aerodynamic response at the respective spanwise position. The local lift fluctuations on the engine/pylon, especially at higher frequencies, have a clear influence on the spectra of the local and global lift coefficient at these frequencies worth considering when evaluating the effects on the aircraft structure. An analysis of the local pressure spectra at different spanwise positions on the wing and the horizontal tailplane allowed a thorough evaluation of the impact of flap track fairings and engine on the local c_p spectra. Thus, conclusions could be drawn from the local c_p spectra to the spanwise load spectra up to the lift spectrum of the entire aircraft. The results demonstrate, that the aerodynamic behavior of the outboard region of the A320 ATRA wing, where no flap track fairings or engines are present, corresponds well with the one of the FNG wing interacting with atmospheric turbulence presented in [12]. By comparing the outboard region of the ATRA wing with the inboard region, the influence of the flow around FTF and engine on the local and global loads was demonstrated. In the vicinity of the flap-track fairing and the engine, it could be shown that the differences in the load fluctuations on the engine geometry for $11 \leq \kappa \leq 28$ are also present in the c_p spectra on the wing's suction side. This demonstrates the influence of the flow around the engine on the aerodynamics on the suction side of the wing at the position of the engine.

Acknowledgements The funding of the activities through the Federal Ministry for Economic Affairs and Climate Action under contracts FKZ 20A1504F in the framework of the air transport research program is gratefully acknowledged. The authors also acknowledge the High Performance Computing Center Stuttgart (HLRS) for the provision of computational resources and the continued support. Special thanks go to Philipp Gansel for the provision of the computational grid.

References

1. European Union Aviation Safety Agency, https://www.easa.europa.eu/downloads/136622/en: Certification Specifications and Acceptable Means of Compliance for Large Aeroplanes CS-25, Amendment 27 (2022). Accessed July 22th, 2022
2. Gansel, P.: HINVA—High-Lift In-Flight Validation TP 3.3 "CFD-Simulation: Flügel/Leitwerk-Interferenz und Trimmung" Projektabschlussbericht I: Einfluss von Flügelnachlauf und Rumpf auf Leitwerksumströmung in Reiseflugkonfiguration. Institute of Aerodynamics and Gas Dynamics, University of Stuttgart (2015). Final report of the research project HINVA (internal)
3. Hancock, P.E., Bradshaw, P.: Turbulence structure of a boundary layer beneath a turbulent free stream. J. Fluid Mech. **205**, 45–76 (1989). https://doi.org/10.1017/S0022112089001941
4. Heinrich, R., Reimer, L.: Comparison of different approaches for gust modeling in the CFD code Tau. Int. Forum Aeroelasticity Struct. Dyn. (2013)
5. Heinrich, R., Reimer, L.: Comparison of different approaches for modeling of atmospheric effects like gusts and wake-vortices in the CFD code Tau. Int. Forum Aeroelasticity Struct. Dyn. (2017)
6. Hoffmann, J.A.: Effects of freestream turbulence on the performance characteristics of an airfoil. AIAA J. **29**(9), 1353–1354 (1991). https://doi.org/10.2514/3.10745
7. Mann, J.: The spatial structure of neutral atmospheric surface-layer turbulence. J. Fluid Mech. **273**, 141–168 (1994). https://doi.org/10.1017/S0022112094001886
8. Mann, J.: Wind field simulation. Probabilistic Eng. Mech. **13**(4), 269–282 (1998). https://doi.org/10.1016/S0266-8920(97)00036-2
9. Müller, J., Ehrle, M., Lutz, T., Krämer, E.: Numerical simulation of the FNG wing section in turbulent inflow. In: Nagel, W.E., Kröner, D.H., Resch, M.M. (eds.), High Performance Computing in Science and Engineering '19, pp. 435–450. Springer, Cham (2021). https://doi.org/10.1007/978-3-030-66792-4_29
10. Müller, J., Hillebrand, M., Lutz, T.: Assessment of the disturbance velocity approach to determine the gust impact on airfoils in transonic flow. In: Dillmann, A., Heller, G., Krämer, E., Wagner, C. (eds.), New Results in Numerical and Experimental Fluid Mechanics, vol. XIII, pp. 14–23. Springer International Publishing, Cham (2021). https://doi.org/10.1007/978-3-030-79561-0_2
11. Müller, J., Hillebrand, M., Schollenberger, M., Ehrle, M., Lutz, T., Krämer, E.: Studies on the applicability of a simplified gust simulation approach in the CFD code TAU. In: Nagel, W.E., Kröner, D.H., Resch, M.M. (eds.), High Performance Computing in Science and Engineering '20, pp. 401–418. Springer, Cham (2021). https://doi.org/10.1007/978-3-030-80602-6_26
12. Müler, J., Lutz, T., Krämer, E.: Numerical Simulation of the Swept FNG Wing in Atmospheric Turbulence. AIAA Aviation 2020 Forum (2020). https://doi.org/10.2514/6.2020-2778

13. Ohno, D., Greiner, M., Müler, J., Rist, U.: Numerical investigations of laminar separation bubbles under the influence of periodic gusts. In: 56th 3AF International Conference on Applied Aerodynamics (2022)
14. Rudnik, R., Reckzeh, D., Quest, J.: HINVA–High lift INflight VAlidation—Project overview and status. In: 50th AIAA Aerospace Sciences Meeting including the New Horizons Forum and Aerospace Exposition (2012). https://doi.org/10.2514/6.2012-106
15. Rudnik, R., Schwetzler, D.: High lift INflight VAlidation (HINVA)—Overview about the 1st flight test Campaign. In: 32nd AIAA Applied Aerodynamics Conference (2014). https://doi.org/10.2514/6.2014-2843
16. Schwamborn, D., Gerhold, T., Heinrich, R.: The DLR TAU-Code, recent applications in research and industry. European Conference on Computational Fluid Dynamics ECCOMAS CFD (2006)
17. Spalart, P.R., Allmaras, S.R.: A one-equation turbulence model for aerodynamic flows. In: 30th Aerospace Sciences Meeting and Exhibit (1992). https://doi.org/10.2514/6.1992-439

Use Case VitAMIn ABC and Subprojects VitAM-Cal and VitAM-Flex

Overview and Results of Subproject VitAMIn-ABC

Kaare Sørensen-Libik

Abstract The VitAMIn-ABC subproject focused on the industrialization of several CFD methods for usage in the aerospace industry. The topics treated included increasing the efficiency of computing large datasets, increasing the accuracy of the turbulence modelling by employing a local approach, the increased automation of the codes to reduce the human effort involved in the data generation, as well as additional activities improving the simulation capabilities of the software. Most of the goals of the project were reached or surpassed, resulting in a considerable impact on the tools and processes in use at Airbus D&S.

Keywords Automatic numerical dataset generation · Turbulence model optimization

1 Introduction

The capabilities of CFD have increased significantly over the last decades. Poweful and accurate methods have been developed and applied to model problems, constantly increasing the fidelity of the physical phenomena simulated and refining the numerical methods defining and solving the resulting discrete systems [1–3]. At the beginning of the VitAM project, a gap between what has been demonstated in a scientific context and what is being employed in industry on a day-to-day basis was however identified, meaning that industry was not taking advantage of the full available potential of numerical methods for aerodynamics. The main goal of the VitAMIn-ABC project has been to close this gap as much as possible, industrializing existing approaches as well as adding new methods that are considered to be of special importance for industrial use. The main effort was directed towards industrializing the aerodynamic parts of the multidisciplinary simulation software, focusing on improving the efficiency and accuracy of the CFD codes already available. In

K. Sørensen-Libik (✉)
Airbus D & S, Rechliner Str., Manching, Germany
e-mail: kaare.sorensen@airbus.com

R. Heinrich (ed.), *Advanced Aircraft Understanding via the Virtual Aircraft Model*, Notes on Numerical Fluid Mechanics and Multidisciplinary Design 155,
https://doi.org/10.1007/978-3-031-69425-7_7

addition, considerable effort was invested into automation of the software suite, with the goal of significantly decreasing the workload of the aerodynamic engineer for a given task, thus increasing the overall efficiency of the aerodynamics group.

2 Improvements in Numerical Efficiency

Since the advent of robust unstructured mesh generators, the very short turn-around time of CFD from receiving a geometry until the first analyses are available has been a strong advantage for industry. Typically, using the in-house unstructured mesher [4, 5], the first computations are completed within a few days of the definition of the aerodynamic shape, including the accurate treatment of engine conditions and the possibility of deploying control surfaces using the Chimera approach [6]. At the beginning of the project, the production of large quantities of data was however not more cost effective than wind tunnel testing, with a typical polar price being similar, or sowhat more expensive than wind tunnel testing, taking into account hardware, energy and human costs in both cases. This under the assumption that non time resolving RANS computations are performed. Due to several methods implemented in the project, both inside the DLR-Tau core solver [7] as well as in the Airbus D&S multidisciplinary software suite, SimServer [8], this cost has been routinely reduced with up to two orders of magnitude for large datasets within the project duration. This has enabled the production of large amounts of high-fidelity data at low cost and within short time periods. Depending on the hardware resources allocated to a given project, typical datasets consiting of several thousand polars can now be generated within a few weeks. This capability is important for several phases of the design life-cycle, starting at the configuration finding phase and ending at datasets for certification. It is now possible to employ highly accurate numerical methods at an early design stage, allowing for making the big decisions of a design on a more reliable and comprehensive basis, reducing the risks of the project. During the various stages of design, full envelope analyses can be integrated in the processes, tracking the impact of local optimizations on the entire flight regime. The datasets, including engine and dynamic effects, can form a basis for the flight dynamics group at early stages of a project, and the local force distributions supplies the loads and structures departments with large amounts of high quality data. Numerical pre-campaigns can be conducted in advance of wind-tunnel campaigns, identifying regions of interest and reducing the costs through pre-definition of store groupings and reduce the testing by basing higher order interaction terms solely on numerical data. The efficiency increase has thus fundamentally changed the way of working not only in the aerodynamics group, and has proven to be a major contribution for faster project development with less risk.

3 Improvements in Turbulence Model Accuracy

The major remaining obstacle for unlocking the full potential of CFD has been the inaccuracies associated with the turbulence modelling. This is especially an issue for high-lift configurations or for low aspect ratio designs typically associated with higly agile aircraft. Significant research has been invested into time-resolving simulation, but thus far these resource intensive computations are not guaranteed to deliver significantly more accurate results than RANS simulation on a regular basis. A major hurdle for RANS turbulence models has been the scope, where the models have been developed with the desire to be able to handle any type of aerodynamic flow within a large range of Reynolds and Mach numbers. This level of generality does however come with a price of reduced accuracy for a particular design where the range of aerodynamic effects and parameters of significance is often limited. Typically, industry has a large amount of experimental data readily available for configurations that are similar to the ongoing design, and more relevant data is typically made available as a design progresses due to the various testing campaigns performed. A natural approach in the effort of increasing modelling accuracy in this context is thus to develop a methodology where the available data can be diectly employed to refine and calibrate the turbulence models used. This requires parametrized turbulence models which can form the basis of some optimization procedure. These parametrized models can be based on pre-existing global models, and it is desirable to introduce terms that have a focused impact on flow phenomena that are considered critical for a given design. Within ViTAMIn-ABC, significant progress was made in this respect, focusing mainly on more localized treatment of vortex flow. These models were then optimized using an automatic procedure for several configurations, resulting in improved accuracy for specific classes of aircraft. The robustness of the optimized models was also analysed by investigating the amount of difference from the calibration case with respect to geometry and flow parameters that would still yield significant modelling improvements. In general, it was observed that the approach would significalty improve the modelling accuracy for variations of a magnitude typically experienced during the development phase of an aircraft, once the topology has been fixed. Separate optimizations would be advantageous if several topolgies are under consideration. Even though the efforts in VitAM on this subject represent the beginning of such an approach in industry, a stage has been reached where the methods can be readily applied to running projects in a robust way, improving the accuracy of the results at a relatively low cost, thus reducing the risks and improving the performance of the product.

4 Improvements in Software Automation

Increased data production rates typically result in additional human workload for the same workflow. In VitAMIn-ABC considerable resources were invested into automating the most labour-intensive parts of the processes, enabling a much higher productivity rate per engineer. Improvements were made in the preparation, computation and postprocessing stages of the workflow, including the assembly of incremental datasets and extraction of local loads. For these stages a typical reduction in human effort of around one order of magnitude was achieved. Other parts of the process, such as geometry preparation and mesh generation were not the focus of the activities in the project, but due to the efficiency improvements elsewhere, these phases have now become more dominant in the overall effort and will be addressed next. Even though there is still potential for increased automation, the progress made in VitAMIn-ABC has resulted in a significant reduction in turn-around times for the generation of large datasets and has considerably reduced the relative workload of the aerodynamic engineer.

5 Improvements in Software Capabilities

In addition to several smaller improvements in the capabilities of the simulation software, some activities in the field of aeroelasticity were also present in VitAMIn-ABC. The project has laid the groundwork for the expansion of the current software environment to efficiently include servo-elastic effects in the datasets.

6 Conclusions

In general, VitAMIn-ABC is considered to have been a very successful project for Airbus D&S, improving the efficiency, accuracy and automation level of the software in use. The cooperation with the involved partners, TU München, TU Braunschweig and DLR was very constructive. A large part of the software developed has been incorporated in the standard toolchain and is in everyday use. The impact of these changes have been profound, resulting in a different way of working not only for the aerodynamics group, but also for the departments requiring aerodynamic input. The improvements are influencing project activities at many stages of the product life, from early design to dataset updates. The unlocking of the full potential of scientific advances in industry is a task that should not be underestimated due to the more stringent requirements of robustness and ease of use in idustrial codes, compared to what is typically accepted for research software. In ViTAMIn-ABC the effort was however proven to be worthwhile, resulting in considerable improvements in quality and efficiency of the work performed in the aerodynamics department.

Acknowledgements The funding of partial activities through the Federal Ministry for Economic Affairs and Climate Action under contracts FKZ 20A1504C in the framework of the air transport research program is gratefully acknowledged.

References

1. Kroll, N., Radespiel, R.V.D. Burg, J.-W., Sørensen, K.: Computational Flight Testing, vol. 123. Springer, NNFM
2. Heinrich, R.: AeroStruct: Enable and Learn How to Integrate Flexibility in Design, vol. 138. Springer, NNFM
3. Kroll, N., Schwamborn, D., Becker, K., Rieger, H., Thiele, F., MEGADESIGN and MegaOpt—German Initiatives for Aerodynamic Simulation and Optimization in Aircraft Design, vol. 107. Springer, NNFM
4. Tremel, U., Sørensen, K.A., Hitzel, S.T., Rieger, H.: Parallel remeshing of unstructured volume grids for CFD applications. IJNMF **53**
5. Tremel, U., Deister, F., Hassan, O., Weatherill, N., Parallel generation of unstructured surface grids. Eng. Comput. **21**
6. Sørensen, K., Tremel, U., Rieger, H., Hitzel, S.T.: Simulation of Manoeuvring Aircraft with the Unstructured Chimera Approach. AIAA 2007-123
7. Gerhold, T., Friedrich, O., Evans, J., Galle, M.: Calculation of Complex Three-dimensional Configurations Employing the DLR-TAU Code. AIAA 97-0167
8. Tremel, U., Deister, F., Sørensen, K.A., Rieger, H., Weatherill, N.P.: The SimServer—A Parallel Multidisciplinary Simulation Environment, DGLS-2003-099

Spalart-Allmaras Turbulence Model Conditioning for Leading-Edge Vortex Flows

Matteo Moioli, Christian Breitsamter, and Kaare Sørensen-Libik

Abstract The physical modeling of turbulence for the partial differential equations closure used to numerically solve large-scale leading-edge vortex flows maintains a significant grade of complexity and interest in the research community. This plays a major factor for the discrepancies with experimental or high fidelity data. A baseline common turbulence model is enhanced by means of a series of additional destruction terms which are formulated with a correlated physical feature of vortical flows as fundament and calibrated by means of an automatic gradient descent optimization procedure towards experimental data. The numerical accuracy is augmented for a certain cluster of cases around the calibration case in the parametric space which describes the classification of vortex flows over highly swept aerodynamic planforms.

Keywords Vortex flow · Turbulence modeling · Optimization · Modeling enhancement

1 Introduction and Motivation

In the design and development phases of an agile and high speed aricraft, the employment of different sources is necessary for a successful mixture of a large amount of data on different fidelity levels. Experimental data from flight test to wind tunnel measurements furnish the highest fidelity but they require the largest need of financial and time resources. The turning time for introducing design variations is also high if not unfeasable. For these reasons, in the last decades the numerical source of data has acquired more and more importance and it still manifests the highest potential for further improvement with respect to both fidelity and cost. In the context of the

M. Moioli (✉) · C. Breitsamter
Chair of Aerodynamics and Fluid Mechanics, Technical University of Munich, Garching, Germany
e-mail: matteo.moioli@aer.mw.tum.de

K. Sørensen-Libik
Airbus Defence and Space, Manching, Germany

R. Heinrich (ed.), *Advanced Aircraft Understanding via the Virtual Aircraft Model*, Notes on Numerical Fluid Mechanics and Multidisciplinary Design 155,
https://doi.org/10.1007/978-3-031-69425-7_8

Advanced Aircraft Understanding via the Virtual Aircraft Model project (VitAM), different research partners have collaborated in order to explore new solutions for the increasing demand in the virtual design phase of an aircraft. A group of partners have established two main directions of improvement have to be investigated with respect to the usage of Computational Fluid Dynamics (CFD) methodologies in the context of aerodynamics cases. These partners are the Technical University of Munich (TUM), the Technical University of Braunschweig (TUBS), and Airbus Defence and Space (ADS). The main problems have been identified in the efficiency and accuracy of the state-of-the-art numerical simulations of aircraft configurations which employs complex flow features such as leading-edge vortex flows. The first issue has been addressed mainly by ADS [1] and the efficiency has been significantly improved by combining a series of available and new numerical methodologies in order to establish a more efficient numerical routine. Afterwards, the main focus of the group has been moved to the modeling of turbulence which is considered to be the main accuracy issue for the numerical simulation of complex aerodynamic cases. The CFD methodologies are commonly based on the finite volume methods formulation and the partial differential equations which solve the physical problem at a cell-level necessitate of a closure equation for the Reynolds stresses computation. At the state of the art, there are different possibilities to close the set of equations where complexity, computational cost, stability and accuracy variate significantly. Until the recent years, the Moore's laws had been able to assure the increasing usage of more complex models which require larger computational costs. However, the expectation of such a predicting law has been recently scaled down [2] due to physical limitations. The possibility of newcoming technologies may provide in the recent future a new trend in augmenting the available computational resources at equal costs. The research group in VitAM has identified the present necessity to enhance the accuracy by maintaining similar computational costs for the most used models of turbulence. These are the turbulence models which base their formulation on the Boussinesq assumption that simplifies the physical meaning of Reynolds stresses. It is assumed that the Reynolds stresses are linearly proportional to the tensor of velocity gradients:

$$-\overline{u_i'u_j'} = 2\nu_t S_{ij} - \frac{2}{3}k\delta_{ij} \tag{1}$$

The Boussinesq assumption is a strong limitation with respect to highly separated or rotational flows where the chaotic nature of the flow is poorly summarized by a linear proportion and alignment of the two tensors. The alternatives to the Boussinesq assumption are several and all promising in relation to their most suited framework of application. With respect to the production of large amounts of data in the context of the establishment of an aerodynamic dataset, eddy viscosity models based on the Boussinesq assumption remain the most used. Moving to more complex turbulence models which solve a larger spectrum of scales introduces significant additional computational costs which makes them unfeasable for very large datasets. Moreover, they encounter the necessity of more complexity in the numerical setup and numerical instabilities [3] without guaranteeing an improvement in accuracy of the results for the complete spectrum of cases [4]. Another possibility with potential is to avoid the limitation of the Boussinesq assumption by employing Reynolds Stress Models (RSM) which increase the computational costs and numerical instabilities in exchange of a higher physical grade of accuracy in the modeling [5]. The authors have decided to concentrate on the eddy viscosity models, in particular to the Spalart-Allmaras model, which is widely used in the context of the numerical simulation of external aerodynamics.

In the VitAM research group, two research directions have been selected. On the one side, TUBS [6] has decided to investigate the possibility to extend the Menter-SST two-equation turbulence model with three additional terms. They are respectively formulated with vortex flows, wake flows and pressure gradient dominated flows as physical fundament. Moreover, a Bayesian Update methodology has been utilized for the calibration in order to include the uncertainty of the target data. On the other side, the authors of this work from TUM together with ADS [7] have followed the direction of enhancing the Spalart-Allmaras one-equation eddy viscosity model for which additional source terms, exclusively active in the vortex field, are added and calibrated by means of a gradient descent algorithm towards experimental data. The idea is to obtain a conditioned model which is employed on a cluster of cases around the calibration target in order to improve the accuracy and not sacrifice the predictive capability on similar test cases.

2 Leading-Edge Vortex Flow

The target flows of the proposed methodology are related to the evolution of large-scale leading-edge vortices. Such kind of flow takes commonly place around highly swept aerodynamic devices like delta wings. For this reason and the large availabilty of experimental data, delta wings are used as test cases for the application of the methods. A delta wing is a triangular wing with a high sweep angle, i.e. larger than 45°. The flow which encounters the high swept leading edge is highly influenced by this geometrical feature together with the angle of attack. In general, for a moderate angle of attack, the flow is not able to remain attached to the leading-edge shape and it separates into a vortex sheet with high momentum and energy which rolls up above the wing surface [8]. It forms a free vortical flow which interacts with the upper wing surface. The vortex consists of an inner viscous sub-core surrounded by a rotational core which is exposed over around 1/3 of the local wing span (Fig. 1). Whereas the sub-core is dominated by viscous phenomena, the rotational core is the one that employs the high tangential velocities which imprints additional suction on the wing surface. The flow outside of the rotational core is pullued around by the transfer of momentum and down towards the wing surface where it determines a reattachment line. The flow which moves from the suction peak outwards encounters a strong adverse pressure gradient which causes a secondary separation into the generation of a secondary vortex. A tertiary structure may also be present with a similar mechanism. Primary and secondary vortices may also interact depending on the flow regime [10],

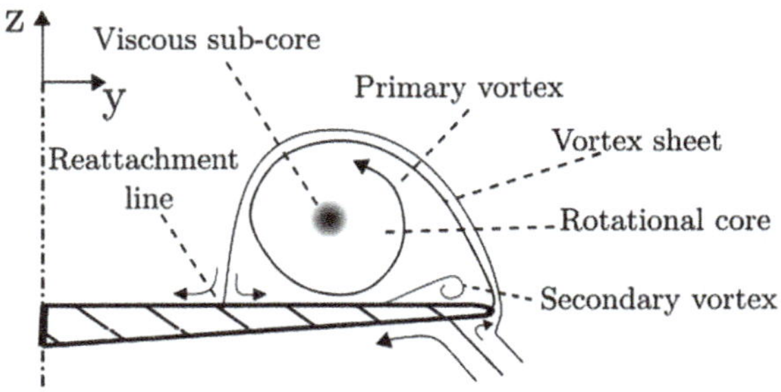

Fig. 1 Schematic representation of a fully developed vortex structure over a delta wing

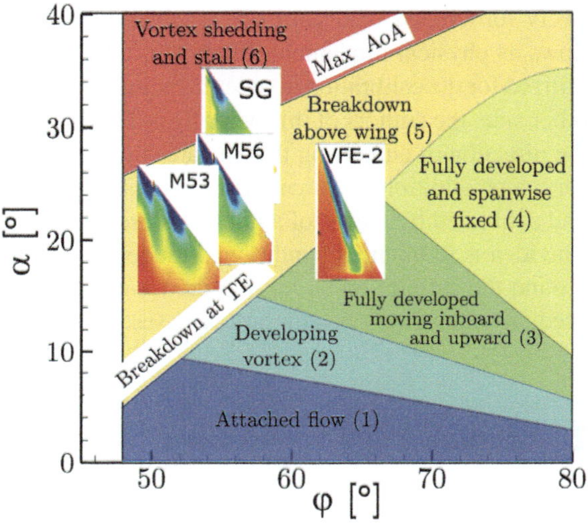

Fig. 2 Mapping of the reported test cases in the parametric space of the angle of attack and the sweep angle which correlate the development stage of a leading-edge vortex flow

the Reynolds number also influences the separation position of partially developed vortices [8]. The vortex suction causes an increase in the wing lift which responds super-linearly to the angle of attack [11]. A vortex developing above a delta wing undergoes a series of states according to its sweep angle and the current angle of attack (Fig. 2). The leading-edge curvature plays an important role at low angles of attack for the separation onset position. In fact, for sharp leading edges, the flow separation is fixed at the wing leading-edge for the whole angle-of-attack envelope. When a small curvature at the leading edge is employed instead, an attached flow is present for low angles of attack (1) (Fig. 2) whereas a partially developed vortex appears by increasing the flow incidence (2). The separation onset moves upstreams until the apex is reached and there it remains fixed like at a sharp leading edge. An additional increase of the angle of attack corresponds to a progressive inward displacement of the vortex structure (3) until it stops and it reaches the stage of a spanwise fixed vortex (4). What happens at even larger angles of attack is of extreme importance for the aerodynamic characteristics of a delta wing and it is notoriously the most problematic flow feature to be properly predicted by the state-of-the-art eddy viscosity models [12, 13]. The vortex reaches a point of instability where its helicity-dominated and structured rotational features collapses into a chaotic and highly turbulent wake flow. The tangential velocities diffuses and the suction distribution loses its peculiar peak shape. The vortex with breakdown stage (5) starts when the breakdown instability takes place at the trailing-edge. Successively, the breakdown migrates upstreams until the vortex sheds directly at the leading edge and a stall condition is achieved (6).

3 Method Description

3.1 Spalart-Allmaras Baseline Model

The baseline turbulence model, used as starting point for the model extension, is the Spalart-Allmaras (SA) one-equation eddy-viscosity model [15] together with the Edwards modifications [16] which improves the near-wall region and the numerical stability.

$$\frac{D\tilde{\nu}}{Dt} = c_{b1}\tilde{S}\tilde{\nu} + \frac{1}{\sigma}\left[\nabla \cdot ((\nu + \tilde{\nu})\nabla\tilde{\nu}) + c_{b2}(\nabla\tilde{\nu})^2\right] - c_{w1} f_w(r) \left[\frac{\tilde{\nu}}{d}\right]^2 . \quad (2)$$

This has been selected as baseline due to its high accuracy with respect to the solution of external aerodynamics cases which employ an attached flow or weak separations. As soon as a highly curved, rotational or separated flow is taken into account, the accuracy drops significantly. Along the years, several modification or variations have been formulated in order to address the lack of accuracy for certain classes of flows. A correction which is commonly employed for vortex dominated flows is the modification of Spalart and Shur [17], which is, however, often too abrupt and its model variation is too drastic for structured vortices and moderate angles of attack. Hence, for several of the state-of-the-art modifications, the focus is concentrated on a certain flow type and they drastically limit the predictive capability in order to improve the accuracy on a flow case.

3.2 Formulation of the Model Augmentation

The proposed methodology [7] focuses on vortex dominated flows, i.e. delta wings, and it has the purpouse to find the best compromise between accuracy and predictive capability by employing an modeling enhancement and an optimization procedure. The objective is to enhance the accuracy for a cluster of cases surrounding the calibration case in the parametric space of the evolution of delta wing vortices (Fig. 2).

3.2.1 Vortex Identification

As first step, a vortex identifier quantity ξ which relates with a vortical structure is formulated (Eq. 3). The quantity is based on the ratio between vorticity and strain rate which is larger than the identity for a vortex flow. In order to switch off the quantity outside of the vortex flow, a limit value c_{vl} is defined to be equal to 1.05. This value arises from the analysis of the ratio inside a boundary layer where it is by good approsimation always lower than 1. Moreover, the limit excludes the other zones of the flow where vorticity is very low. The objective here is to use the ξ for every additional term included in the model enhancement to avoid that the model variation provokes a modification of the modeling of turbulence in the regions of the flow where the baseline is accurate, i.e. the boundary layer and attached flows. A low value ϵ is used to avoid overflow errors if the strain rate is locally 0.

$$\xi = max\left[\left(\left(\frac{\omega}{S+\epsilon}\right) - c_{vl}\right), 0.0\right] \tag{3}$$

3.2.2 Terms Formulation

All the terms introduced in the model modification are formulated in a form similar to the baseline production term but associated to variables and functions which relate to physical features of different regions or types of vortices. In this way, the model potential accuracy is enhanced by means of its flexibility to target vortex flows in their physical detail by means of an automatic calibration. The additional terms (Eq. 4) are distinguished into two main sub-groups: the one (r) which targets different radial-wise influences and the second (a) which focuses on the different flow types encountered along the vortex-axis development.

$$\begin{aligned}
\frac{D\tilde{\nu}}{Dt} = c_{b1}\tilde{S}(1-f_{t2})\tilde{\nu} - \Bigg[& \overbrace{c_{bv1}\xi S\tilde{\nu}}^{(r.1)} + \overbrace{c_{bv2}\xi^{\frac{1}{2}} S\tilde{\nu}}^{(r.2)} + \overbrace{c_{bv3}\xi^{2} S\tilde{\nu}}^{(r.3)} \\
& + \overbrace{c_{bv4}\left(min\left(\frac{1}{\xi}, c_{vr,lim}\right)\right)\xi_{sw} S\tilde{\nu}}^{(r.4)} + \overbrace{c_{bvh1}\xi\widetilde{\mathcal{H}}\tilde{\nu}}^{(a.1)} \\
& + \overbrace{c_{bvh2}\left[max\left(min\left(\frac{1}{\widetilde{\mathcal{H}}}, 1.0\right), 0.0\right)\right]\xi\omega\tilde{\nu}}^{(a.2)} + \overbrace{c_{bvb}\xi||\nabla\mathbf{V}\frac{\omega}{||\omega||}||\tilde{\nu}}^{(a.3)} \Bigg] \\
& + \frac{1}{\sigma}\left[\nabla\cdot((\nu+\tilde{\nu})\nabla\tilde{\nu}) + c_{b2}(\nabla\tilde{\nu})^2\right] - (c_{w1}f_w - \frac{c_{b1}}{\kappa^2}f_{t2})\left[\frac{\tilde{\nu}}{d}\right]^2
\end{aligned} \tag{4}$$

With

$$\mathcal{H} = V \cdot \omega \tag{5}$$

$$\tilde{\mathcal{H}} = \frac{|\mathcal{H}|}{U_\infty} \tag{6}$$

$$\xi_{sw} = \frac{\xi}{\xi + \epsilon} \tag{7}$$

The ξ quantity is not used exclusively as a switch factor but its distribution inside the vortex region is employed to distinguish different radial-wise distributions in the first sub-group. By means of different exponentiation laws of ξ, a series of terms are formulated which are based on the linear (r.1), sublinear (r.2), superlinear (r.3) and inverse exponentiation to ξ (r.4). The second group is motivated by the importance of the breakdown instability and the abrupt change of the flow upstream and downstream of it. The jet-type vortex and highly structured vortex present upstream of breakdown is highlighted by the alignment of axial velocity and axial vorticity. This physical content is properly described by the helicity $\mathcal{H}$. In order to avoid the representation of the vortex chirality and the change of the term dimensionality, a non-dimensional and invariant form $\widetilde{\mathcal{H}}$ of helicity is defined and applied to the first term (a.1). With respect to the wake flow downstream of breakdown, a term (a.2) is associated with a switch function based on the inverse of $\widetilde{\mathcal{H}}$. The proximity of breakdown is highlighted by the last term (a.3) which is based on the variation of the alignment of the velocity gradient tensor and the vorticity direction.

3.3 *Optimization*

The optimization procedure is performed by means of a gradient descent algorithm which minimizes the mean absolute error between the experimental data and the numerical results. The toolchain handles both the algorithm computations, the run of the numerical simulations and the I/O between the two compartments. The gradient descent permits to obtain a fast convergence but it is inherently local. Therefore, the experience from previous sensitivity tests or other test cases is important to target meaningful directions and select starting points of the coefficients. To overcome the locality, different optimization procedures from different starting points are performed in order to verify the best possible optimum. When the optimum is identified, it is applied to a larger set of cases for the considered calibration test cases, e.g. the whole angle-of-attack polar, and the accuracy enhancement is assessed. Successively, the predictive capability is investigated with the application of the optimum to one or more test cases which employ mild variations from the calibration case in terms of their geometrical or flow condition parameters. Therefore, the potential of the optimal coefficients to improve a cluster of cases around the calibrated case is investigated.

4 Results

4.1 Numerical Setup

The numerical simulations and the method implementation are performed with the TAU-Code [14] which has been developed by the German Aerospace Center (DLR). It solves a cell-vertex formulation of the finite volume discretization and it permits the usage of hybrid-unstructured numerical grids which have the capability to properly solve the near-weall region while being flexible enough for the discretization of complex geometries. The solver is stabilized with an implicit dual-time stepping and a multigrid relaxation methodology is used. A LUSGS linear solver is utilized in combination with a Backward-Euler method. The flux discretization is performed with an AUSMDV method that is a hybrid formulation of the AUSM (Advection Upstream Splitting Method). It merges the Difference (AUSMD) and the Vector splitting (AUSMV) formulations. All the numerical simulations are three-dimensional cases, share a compressible fully-turbulent flow and employ the SAE turbulence model and/or the optimized version formulated in this research work.

4.2 65° Swept Delta Wing: VFE-2

A first section of the numerical results is dedicated to the 65° swept delta wing which has been deeply investigated by the NASA experiments [18] to the collaboration of different partners in the Second Vortex Flow Experiment (VFE-2) campaign [19]. The experimental data are available for a large number of flow conditions and geometrical variations. This and the fact that it has been numerically investigated by several international partners [20] makes it a perfect starting test case for the application of the proposed methodology. The SA model related optimization is performed on the configuration at subsonic conditions and with a leading-edge radius equal to 0.05% of l_μ [7] which forces the vortex separation onset at the wing apex without encountering partially developed vortex flow stages. The main source of discrepancy on this configuration is the breakdown location and the C_P distribution downstream and on the rear part of the wing. The turbulence model modification (Table 1) is able to target this source of error from the experiments. The results on the calibration case manifests the best overall improvement in terms of accuracy [7]. In order to investigate the predictive capability extension of the model for a cluster of cases around the calibrtion target, the optimum is also applied to a series of different

Table 1 Optimal coefficients calibration [7] for the 65° swept delta wing

Coefficient	c_{bv1}	c_{bv4}	c_{bvh1}	c_{bvh2}	c_{bvb}
Value	0.0375	0.003	−0.02	−0.432	−0.6

Table 2 Mean absolute error (MAE) of C_P of the validation cases for the optimum turbulence model coefficient set on the 65° swept delta wing with a leading edge radius of 0.05% of l_μ

LE radius (% of l_μ)	Re	MAE_{SAE}	MAE_{Opt}	$(MAE_{SAE} - MAE_{Opt})$ / MAE_{SAE} (%)
0.0	$2 \cdot 10^6$	0.1355	0.1195	−11.8%
0.0	$6 \cdot 10^6$	0.1349	0.0985	−26.9%
0.05	$60 \cdot 10^6$	0.1831	0.1452	−20.6%
0.15	$2 \cdot 10^6$	0.14577	0.1276	−12.5%
0.15	$3 \cdot 10^6$	0.1934	0.1745	−10%
0.15	$6 \cdot 10^6$	0.1177	0.1150	−2.2%
Mean	–	0.1517	0.1300	−14.3%

flow conditions and geometrical variations (Table 2) for which experimental data are available. It is evident, the more the validation case differs from the original calibration case, the less accuracy enhancement is achieved with the same calibration. However, a significant improvement in terms of accuracy is present for all the cases except for the larger leading-edge radius of 0.15% of l_μ. On the one hand, in this case, the accuracy of the baseline model is already appropriate for the largest Reynolds number of $6 \cdot 10^6$ and the low variation is an indication of the well behavior of the model enhacement which does not alter already well predicted conditions and vortex flows. On the other hand, the lower Reynolds number cases at this leading-edge curvature manifest larger errors in particular for $Re = 3 \cdot 10^6$. This indicates that, due to the small leading-edge curvature, a different development of the vortex flow stage takes place and the sensitivity to Reynolds number increases together with the importance of the vortex separation position and type. The angle of attack around 20° is reported in Fig. 3 where the calibration and a validation case—with a sharp leading edge and a Reynolds number of $3 \cdot 10^6$—are shown by means of cross-flow cuts at the locations with the highest sensitivity to turbulence modeling. It is evident that the calibration model shows the highest accuracy enhancement. In fact, the optimized model is able to match the experimental data with a high accuracy. The breakdown location is displaced downstream compared to the baseline model and the suction peak level is in agreement with the experiments. Also the pressure distribution associated with the secondary vortex and the suction distribution outboard are closer to the experimental data. The validation case shows that the vortex flow stage is similar but its evolution and suction distribution changes as consequence of the different leading-edge curvature and the Reynolds number. This has the consequence that the optimum is still able to change the vortex flow stage in the right direction but it overpredicts the necessary correction of the eddy viscosity destruction in the vortex flow field. As a consequence, the vortex stage is improved but the C_P distribution shows some offset to the experimental data with respect to the vortex suction peak. Overall, the improvement on the full set of validation cases is significant and it shows the potential of the methodology application for both the objectives to increase the

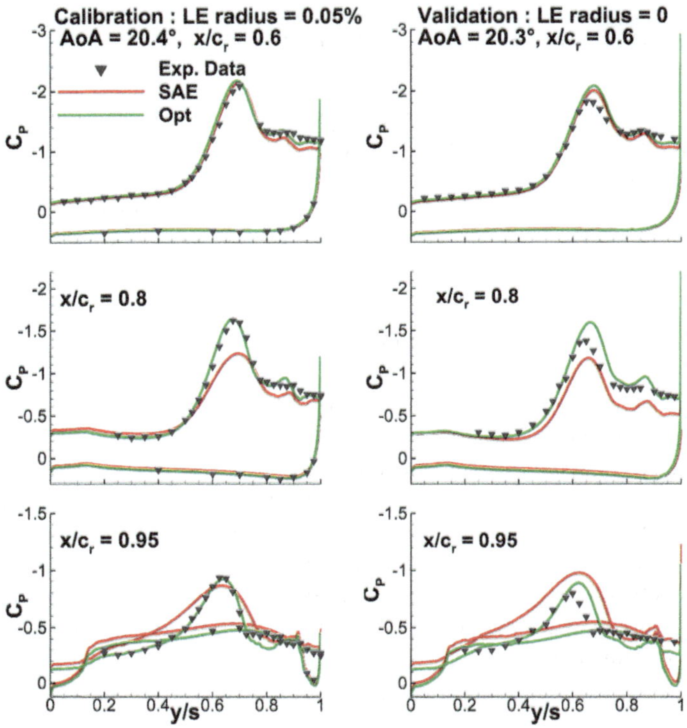

Fig. 3 The C_P distribution for cross-flow sections is shown for the calibration case [7] and a validation case. The experimental data are compared to the baseline SAE and the optimized model (Table 1)

accuracy towards experiments and the predictive capability for a cluster of cases with mild variation from the calibration case.

4.3 Moderate Swept Delta Wings: Calibration on the Model53 and Validation

The following test case has been established in order to answer two fundamental requirements of the research which are, first, the need of experimental data for extending the calibration panorama and, second, the investigation of the methodology potential for a more complex class of vortex flow cases. Therefore, as a first step, a novel manufacturing method has been developed in order to rapidly prototype low-subsonic wind tunnel models by means of the fused deposition of polyactide material [21] with integrated pressure measurement taps. Using this procedure, two models are produced which consist of two sweep angle variations, i.e. 53° and 56°, of the same delta wing planform with a 20° deployed leading-edge slat and a generic peniche and fuselage part. As a second step, the two wind tunnel models are measured at the W/T-A facility of TUM-AER and the experimental data are employed for the application of the present methodology. The Model53 is used as calibration target whereas the Model56 is employed for the validation of the previously optimized set of turbulence model terms. In addition, another two validation cases with larger geometrical and flow condition differences from the calibration are briefly reported for the most significant angles of attack. These two additional cases are the AVT-183 53° diamond wing [22] and a 55° swept blended wing body [23].

4.3.1 Model53 Versus Model56: Calibration and Sweep Angle Variation

After a first analysis of the numerical results provided by the baseline SAE model and its discrepancies to experimental data, the main target of the calibration has been set to the breakdown characteristics, strength and location/presence of which are related to the correct prediction of the stall angle of attack. The experimental data set as an optimization target is the one at 28° which shows the most significant difference to the baseline model and they regroup the main differences seen for the other high angles of attack. The optimization procedure returns an optimal set of coefficients (Table 3) which is then applied to the range of large angles of attack for both calibration and validation cases. The average improvement of the mean absolute error between the numerical and experimental pressure distribution is of 37.46% for the Model53 and 39.3% for the Model56. This result indicates that almost no accuracy improvement drops with a relatively small leading-edge sweep variation of the same planform and it shows the potential of the predictive capability of the methodology. Figure 4 shows the improvement for the two flow angles where the main discrepancy with SAE is present. The discrepancy is mainly related to the correct displacement upstream of the breakdown position until the stall condition is reached. The improvement is significant for both cases. In fact, the pressure distribution loses the curved distribution, peculiar of a vortex footprint, and it acquires a flat distribution at a lower suction level for the whole wing extension. The evolution of the suction along the flow direction properly matches the experiments in the first half of the wing whereas on the rear part a certain offset is present. In general, the mean error improvement is significant. Also, the experimental data available for the pressure side of the wing show how the methodology succeeds in avoiding sensitivity in the flow regions which are already well calibrated by the baseline model, i.e. a fully attached flow.

Table 3 Calibration optimum for the Model53

Coefficient	c_{bv1}	c_{bvh1}	c_{bvh2}
Value	0.41	−0.11	−0.10

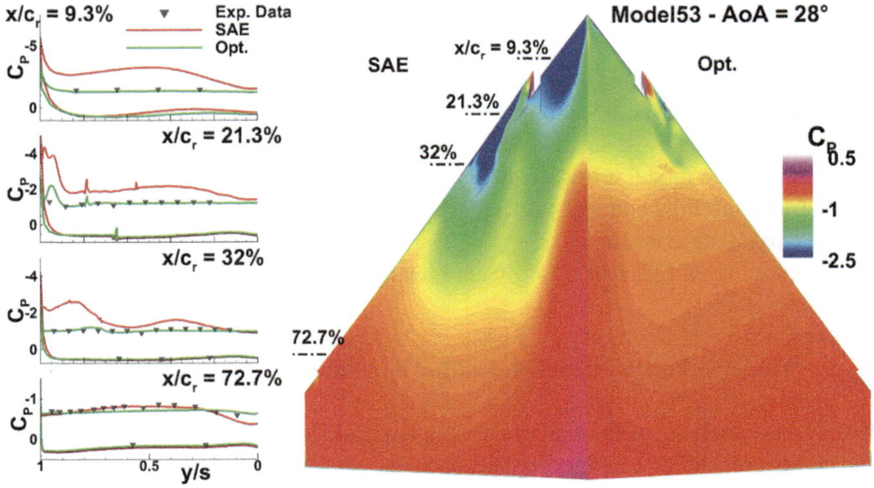

(a) Model53 - Calibration case, AoA = 28°.

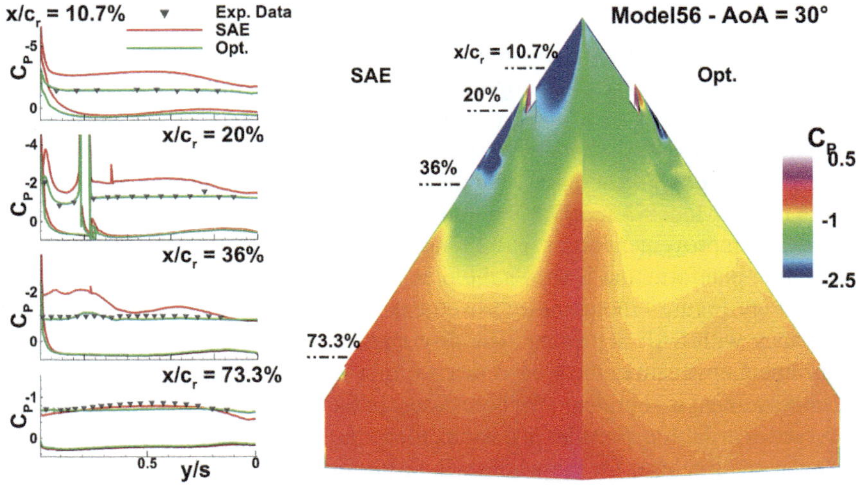

(b) Model56 - Validation case, AoA = 30°.

Fig. 4 The upper surface C_P contour plot and 4 cross-flow pressure sections are reported for the stall condition of the Model53 and Model56. The original SAE model and the optimized turbulence model (Table 3) are compared to the experimental data

4.3.2 Validation on the AVT-183 53° Diamond Wing and the 55° Blended Wing Body

As introduced at the beginning of this section, an objective of the research is to find the potential extension of the predictive capability of the proposed methodology. The most trivial option is to parametrize the different test cases which have been investigated (Fig. 2) and identify clusters of cases which share similarities with respect to important aspects of the vortex flow development. In this case, the analysis of the well behavior of the Model53 calibration is analyzed with respect to two different test cases which are the AVT-183 diamond wing and the 55° blended wing body. They share a similar sweep angle and flow condition but they differs with respect to several geometrical parameters like the body thickness, the leading-edge roundness and the trailing-edge sweep. The application of the Model53 optimum on the range of high angles of attack manifests the good behavior of the methodology to target a class of flow conditions on a test case and being able to translate the improvement to similar test cases. Figure 5 shows how the vortex flow stage is modified with the application of the optimized turbulence model. At the section in proximity of the wing apex at $x/c_r = 0.1$ a clear difference is present between the two test cases and this shows how the difference in terms of geometrical parameters are important with respect to the vortex evolution. Nevertherless, the same optimal set of coefficients is able to significantly improve the pressure distribution. The diamond wing manifests a completely stalled flow and the match with experiments is quantitatively and qualitatively appropriate. The blended wing body shows the presence of a vortical structure and the optimum is able to maintain a similar qualitative distribution but it improves the suction level. The improvement is consistent moving downstream to $x/c_r = 0.2$ but a certain discrepancy is present between the optimzed results and experiments. The new model overpredicts the suction drop along the wing surface below the wake flow. This determins a slight reduction in accuracy compared to the baseline model but it remains appropriate. Overall, the improvemet is consistent with which has been seen on the calibration case. Therefore, the results show the potential of the approach to increase the accuracy and maintain a suitable predictive capability for a relative large cluster of cases around the calibration target.

5 Conclusion

A research partnership has been established in the context of the Advanced Aircraft Understanding via the Virtual Aircraft Model project in order to investigate the possibility to enhance efficiency and accuracy of state-of-the-art numerical simulations of vortex dominated flows while maintaining appropriate computational costs and predictive capabilities. The methodology proposed by the Technical University of Munich extends the one-equation Spalart Allmaras eddy viscosity model to vortex flows by the addition of properly formulated production terms which is exclusively

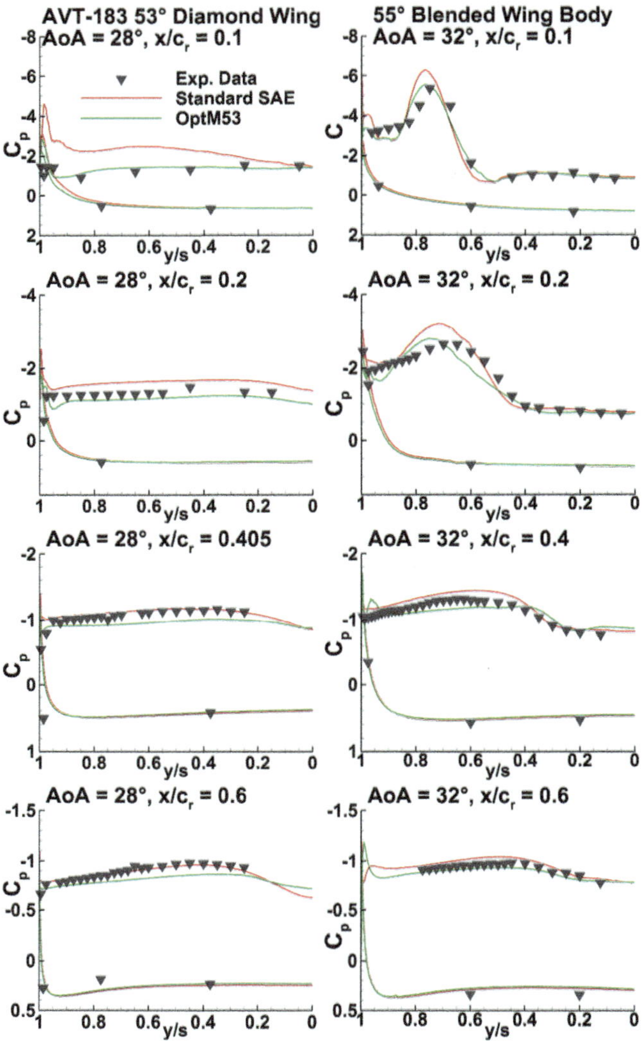

Fig. 5 4 cross-flow pressure sections are reported for the validation cases, i.e. the AVT-183 diamond wing and the 55° blended wing body near the stall condition. The original SAE model and the optimized turbulence model for the Model53 (Table 3) are compared to the experimental data

activated in the vortical region without compromising the original calibration for fully attached flows. The extended model is coupled with an optimization procedure towards experimental data. The derived optimum is applied to calibration cases at different kinds and characteristics of vortex flows to show its accuracy improvement potential. Afterwards, it is applied to validation cases which cover a cluster around the calibration target where geometrical and flow conditions mildly variate. The predictive capability for the application to a large data-set of cases with mild to moderate variations is confirmed as well as the proper numerical stability and computational costs. The model and methodology may be further extended in terms of the introduction of new terms, extensions, new optimization techniques, prediction methods as well as its transposition to other baseline turbulence models.

Acknowledgements The support to this research work by Airbus Defence and Space in the LUFO V-2 project VitAM/VitaMInABC (Virtual Aircraft Model for the Industrial Assessment of Blended Wing Body Controllability, FKZ: 20A1504C), which has been partially funded by the German Federal Ministry for Economic Affairs and Energy (BMWi), is gratefully acknowledged. The authors thank the German Aerospace Center (DLR) for the permission to use the DLR TAU-Code to perform the methodology implementation and the numerical investigations. The authors gratefully acknowledge the Gauss Centre for Supercomputing e.V. (www.gauss-centre.eu) for providing computing time on the GCS Supercomputer SuperMUC at Leibniz Supercomputing Center (LRZ, www.lrz.de) in order to successfully perform the research tasks.

References

1. Sørensen, K.A.: VitAMin-ABC, Overview of Airbus D and S activities in VitAM (2018)
2. Schuster, P.: The end of Moore's law: living without an exponential increase in the efficiency of computational facilities. Complexity **21**(S1), 6–9 (2016). https://doi.org/10.1002/cplx.21824
3. Basara, B., Jakirlic, S.: A new hybrid turbulence modelling strategy for industrial CFD. Int. J. Numer. Meth. Fluids **42**(1), 89–116 (2003). https://doi.org/10.1002/fld.492
4. Freeman, J.: Computational Fluid Dynamics Investigation of Vortex Breakdown for a Delta Wing at High Angle of Attack, Thesis, Department of the Air Force University, Air Force Institute of Technology, USAF, pp. 172 (2003). https://doi.org/10.1002/cplx.21824
5. Eisfeld, B., Rumsey, C., Togiti, V.: Verification and validation of a second-moment-closure model. AIAA J. **54**(5), 1524–1541 (2016). https://doi.org/10.2514/1.J054718
6. Gokul, S., de Souza, A.C.B., Radespiel, R., Zander, E., Friedman, N., Moshagen, T., Moioli, M., Breitsamter, C., Sørensen, K.A.: Calibration of an extended Eddy Viscosity Turbulence Model using Uncertainty Quantification, AIAA Scitech 2020 Forum (2020). https://doi.org/10.2514/6.2020-1031
7. Moioli, M., Breitsamter, C., Sørensen, K.A.: Parametric data-based turbulence modelling for vortex dominated flows. Int. J. Comput. Fluid Dyn., Taylor Fr. **33**(4), 149–170 (2019). https://doi.org/10.1080/10618562.2019.1617857
8. Hummel, D.: Experimental investigation of the flow on the suction side of a thin delta wing. NASA Tech. Memo. **75897** (1981)
9. Breitsamter, C.: Unsteady flow phenomena associated with leading-edge vortices. Prog. Aerosp. Sci. **44**(1), 48–65 (2008). https://doi.org/10.1016/j.paerosci.2007.10.002
10. Hummel, D.: On the Vortex Formation over a Slender Wing at Larg Angles of Incidence, High Angle of Attack Aerodynamics, CP-247, AGARD, pp. 15-1–15-17 (1979)
11. Polhamus, E.C.: Predictions of vortex-lift characteristics by a leading-edge suctionanalogy. J. Aircr. **8**(4), 193–199 (1971). https://doi.org/10.2514/3.44254
12. LeRoy, J., Mary, I. and Rodriguez, O.: CFD Solutions of 70-deg Delta Wing Flows, 21st AIAA Applied Aerodynamics Conference (2003). https://doi.org/10.2514/6.2003-4219
13. Cummings, R.M., Forsythe, J.R., Morton, S.A., Squires, K.D.: Computational challenges in high angle of attack flow prediction. Prog. Aerosp. Sci. **39**(5), 369–384 (2003). https://doi.org/10.1016/S0376-0421(03)00041-1
14. Gerhold, T.: Overview of the Hybrid RANS Code TAU, MEGAFLOW—Numerical Flow Simulation for Aircraft Design. Springer, Berlin, Heidelberg, pp. 81–92 (2005). isbn 978-3-540-32382-2

15. Spalart, P. and Allmaras, S.: A One-Equation Turbulence Model for Aerodynamic Flows, AIAA, vol. 439 (1992). https://doi.org/10.2514/6.1992-439
16. Edwards, J.R., Chandra, S.: Comparison of eddy viscosity-transport turbulence models for three-dimensional, shock-separated flowfields. AIAA J. **34**(4), 756–763 (1996). https://doi.org/10.2514/3.13137
17. Shur, M.L., Strelets, M.K., Travin, A.K., Spalart, P.R.: Turbulence modeling in rotating and curved channels: assessing the spalart-shur correction. AIAA J. **38**(5), 784–792 (2000). https://doi.org/10.2514/2.1058
18. Chu, J., Luckring, J.M.: Experimental surface pressure data obtained on 65° delta wing across reynolds number and mach number ranges. NASA Technical Memorandum, vol. 4645. Langley Research Center, Hampton, Virginia (1996)
19. Luckring, J.M., Hummel, D.: What Was Learned from the New VFE–2 Experiments? In: 46th AIAA Aerospace Sciences Meeting and Exhibit, Reno, NV (2008). https://doi.org/10.2514/6.2008-383
20. Fritz, W., Cummings, R.M.: What was Learned from the Numerical Simulations for the VFE–2? AIAA Paper 2008–0399 (2008). https://doi.org/10.2514/6.2008-399
21. Moioli, M., Reinbold, C., Sørensen, K., Breitsamter, C.: Investigation of additively manufactured wind tunnel models with integrated pressure taps for vortex flow analysis. Aerospace **6**, 113 (2019). https://doi.org/10.3390/aerospace6100113
22. Luckring, J.M., Boelens, O.J., Breitsamter, C., Hövelmann, A., Knoth, F., Malloy, D.J., Deck, S.: Objectives, approach, and scope for the AVT-183 diamond-wing investigations. Aerosp. Sci. Technol. **57**, 2–17 (2016). https://doi.org/10.1016/j.ast.2016.05.025
23. Hövelmann, A., Breitsamter, C.: Leading-Edge geometry effects on the vortex formation of a diamond-wing configuration. J. Aircr. **52**(5), 1596–1610 (2015). https://doi.org/10.2514/1.C033014

Bayesian Calibration of Extended Menter-SST Turbulence Model for Vortical Flows

Gokul Subbian, Thilo Moshagen, Noemi Friedman, Elmar Zander, and Rolf Radespiel

Abstract Vortical flow fields are complex in nature and are prevalent in aircraft aerodynamics. Reynolds Averaged Navier Stokes simulation methodology along with eddy viscosity based Menter—Shear Stress Transport model is widely used in the research community for practical applications. Extensions to Menter—SST turbulence model based on flow physics are explored in the present work to improve the prediction capability for separated and vortical flow fields. A data driven calibration approach is employed based on Bayesian modeling to determine the coefficients of the extended turbulence model. Performance of the extended model is studied with complex delta wings under subsonic flow conditions.

Keywords Vortical flow · Turbulence modeling · Calibration using Bayesian update

1 Introduction

Mathematical models along with the assistance of hardware computing technologies play a major role in the design and development of aircraft. Numerical tools are carefully selected for the problem to be solved, based on the availability of required

---

G. Subbian (✉) · R. Radespiel
Institute of Fluid Mechanics, Hermann-Blenk-Straße 37, 38108 Braunschweig, Germany
e-mail: g.subbian@tu-braunschweig.de

R. Radespiel
e-mail: r.radespiel@tu-braunschweig.de

T. Moshagen
Maschinenbau/Verfahrens- und Umwelttechnik, Hochschule Wismar, Germany

N. Friedman
Institute for Computer Science and Control (SZTAKI), Budapest, Hungary

E. Zander
Institute of Scientific Computing, TU Braunschweig, Braunschweig, Germany

R. Heinrich (ed.), *Advanced Aircraft Understanding via the Virtual Aircraft Model*, Notes on Numerical Fluid Mechanics and Multidisciplinary Design 155,
https://doi.org/10.1007/978-3-031-69425-7_9

resources. The process of an aerodynamic geometry to be proposed, designed and tested in wind tunnel is resource intensive with highest level of accuracy and performance expectations. Computer aided simulations pitch in during the initial design process, offering more flexibility to explore the design space and its consequent outcomes. The purpose of the VIrTual Aircraft Model (VITAM) project is to understand and redefine the scope of these numerical tools and push the limitations of these methods to the next level. In the task group of the VITAM–Cal (Aircraft Model Calibration) two different directions were explored for model improvements and calibration. Extensions to two equation eddy viscosity based Menter–Shear Stress Transport (SST) turbulence model were explored and Bayesian update based on Uncertainty Quantification was used for the calibration process, by Technische Universität Braunschweig. Another approach with one equation eddy viscosity based Spalart Allmaras turbulence model and Deterministic Optimization was pursued by Technical University of Munich, in parallel [1].

1.1 Motivation

Different analogies exist for numerical simulations of fluid flows. In particular, Reynolds Averaged Navier Stokes (RANS) technique is one of the common simulation tool employed by the research community. Primary reason being the computational cost effectiveness without a large compromise on the accuracy of simulated results. This also depends on the underlying turbulence model used during the RANS simulations. Similar to the popularity of the RANS approach, two equation eddy viscosity based Menter—SST turbulence model is widely used due to its robustness, resource requirement and prediction capabilities. External aircraft aerodynamics is the primary area of interest and a major part of it is covered well by this approach. However, shortcomings of this modeling approach are known to exist in complex flow fields such as separated and vortical flows. In the case of low aspect ratio wing configurations, such as delta wings, lift generated by vortical flow over the wing is very important. Vortex breakdown over the wing surfaces of delta wings have a critical effect on the pitching moment characteristics. Furthermore, with respect to transport aircraft wings with high lift devices, nacelle strakes in particular are incorporated to improve the lift characteristics. Therefore, it is important to accurately predict and understand these flow fields. Different approaches are available, with the view to improve the prediction capabilities of the relatively simple but effective Menter—SST turbulence model. This paper deals with three different flow-physics based extensions, which are incorporated to the original Menter—SST turbulence model. Any new turbulence model should be calibrated, verified and validated, generally using wind tunnel results, before it could be employed for numerical predictions. Calibration process is usually performed by a research engineer in a manual process and it is relatively a tedious and complex process. Currently large data sets of numerical simulations are produced by high performance computing clusters and these data sets are an asset to data driven modeling approaches. For the extended Menter—SST turbulence model developed during the VITAM project, one such novel approach of Bayesian modeling is used to calibrate and determine the extended turbulence model coefficients. Finally, the model is verified and validated using generic test cases, along with a new complex delta wing configuration, which serves as the target test case.

2 Numerical Model

In this section, an overview of the extensions to the Menter—SST turbulence model is provided.

2.1 Menter—SST Turbulence Model

Menter—Shear Stress Transport turbulence model is a combination of two two-equation turbulence models, namely k-ω model and k-ϵ model [15]. Near wall region is modeled using k-ω and in the freestream k-ϵ model is employed with the usage of a blending function. This model is well known and so the two transport equations are given here.

$$\frac{D\rho k}{Dt} = \tau_{ij}\frac{\partial u_i}{\partial x_j} - \beta^* \rho \omega k + \frac{\partial}{\partial x_j}\left[(\mu + \sigma_k \mu_t)\frac{\partial k}{\partial x_j}\right] \tag{1}$$

$$\frac{D\rho\omega}{Dt} = \frac{\gamma}{\nu_t}\tau_{ij}\frac{\partial u_i}{\partial x_j} - \beta\rho\omega^2 + \frac{\partial}{\partial x_j}\left[(\mu + \sigma_\omega \mu_t)\frac{\partial \omega}{\partial x_j}\right] + 2\left(1 - F_1\right)\rho\sigma_{\omega 2}\frac{1}{\omega}\frac{\partial k}{\partial x_j}\frac{\partial \omega}{\partial x_j} \tag{2}$$

Boussinesq assumption is used to calculate the turbulent reynolds stresses using the eddy viscosity.

$$-\widetilde{\overline{\rho} u_i'' u_j''} = \mu_t\left(\frac{\partial \tilde{u_i}}{\partial x_j} + \frac{\partial \tilde{u_j}}{\partial x_i} - \frac{2}{3}\frac{\partial \tilde{u_k}}{\partial x_k}\delta_{ij}\right) - \frac{2}{3}\overline{\rho} k \delta_{ij} \tag{3}$$

2.2 Spalart Allmaras Based Rotation and Curvature Correction

The first physical extension is to improve the Menter—SST turbulence model for flows involving system rotation and streamline curvatures. The term was originally proposed by Spalart—Shur [3], based on the previous work of Knight and Saffman's on the "gyroscopic stability effect" term [2]. The gyroscopic stability term includes the angular velocity of the principal axes of the mean rate of strain tensor, introduced to the transport equations. On this basis, Spalart—Shur developed the Spalart Allmaras based Rotation and Curvature correction (SARC) formula [3], which is Galilean invariant [5] and is added to the Spalart—Allmaras one equation turbulence model.

$$f_{rotation} = (1 + c_{r1}) \frac{2r^*}{1+r^*} \left[1 - c_{r3} tan^{-1} (c_{r2}\tilde{r})\right] - c_{r1} \quad (4)$$

There exists a difference in the production terms between SA and Menter—SST model, as SA model production terms are based on the vorticity tensor Ω whereas Menter—SST model production terms are based on the strain rate tensor S [4]. Menter adapted this original SARC term (4) by introducing limiters to limit production as well as for numerical stability and added it to the Menter—SST model [4].

$$f_{r1} = max \{min (f_{rotation}, 1.25), 0.0\} \quad (5)$$

Menter—SST model with SARC term is given by [4]:

$$\frac{\partial(\rho\omega)}{\partial t} + \frac{\partial(\rho u_j \omega)}{\partial x_j} = \alpha \frac{\rho P_k}{\mu_t} f_{r1} - D_\omega + Cd_\omega + \frac{\partial}{\partial x_j} \left[\mu_{ef} \frac{\partial \omega}{\partial x_j}\right] \quad (6)$$

The values of the correction function constants c_{r1}, c_{r2} and c_{r3} as proposed by Menter are 1, 2 and 1 respectively.

2.3 Scale Adaptive Simulation Based Correction

The second extension deals with sensitizing Menter—SST model for largely separated flows and flows with highly curved velocity profiles. This can be regulated by increasing the turbulence production or decreasing the dissipation levels, since lower levels of Reynolds stresses prediction by Menter—SST model is observed in these flow conditions. Menter derived a Scale-Adaptive-Simulation (SAS) based model, which includes the mean—flow inhomogeneity [6]. The SAS model term developed my Menter is based on the $k - kL$ model originated by Rotta, and it is used as a source term scaled with the second derivative of the mean flow added to the ω length scale equation. Improvements over standard model have already been shown in the literature [6]. This approach has been adapted in the development of Reynolds Stress Models (RSM) by Jakirlić and Maduta [7] and Cecora et al. [8]. Same process as in RSM model implementation has been adapted to include the term with the Menter—SST model. SAS model is only the naming convention and no scale resolving simulations are performed in the present work. Based on the coefficient determined by the calibration process, the SAS term acts either as a source or sink in the length—scale equation.

$$Q_{SAS} = max \left[\rho \zeta_2 \kappa S^2 \left(\frac{L}{L_{vk}}\right)^2 - C \cdot \frac{2\rho k}{\sigma_\Phi} max \left(\frac{|\nabla\omega|^2}{\omega^2}, \frac{|\nabla k|^2}{k^2}\right), 0\right] \quad (7)$$

This Q_{SAS} term (Eq. 7) is initially added as a sink term in Eq. 2. The length scale ratios in Eq. 7 are given by:

$$L = \frac{\sqrt{k}}{\left(c_\mu^{1/4} \cdot \omega\right)} \tag{8}$$

$$L_{vk} = \frac{\kappa S}{|\nabla^2 U|} \tag{9}$$

2.4 *Irrotational Strains*

The third extension deals with sensitizing the Menter—SST model in flows with separation induced by adverse pressure gradient. Probst [9] derived a source term for a Reynolds Stress Model, accounting for adverse pressure gradients in the flowfield, based on the work of Hanjalić and Launder to sensitize the k-ϵ turbulence model's dissipation equation to irrotational strains [10]. The term is based on a streamline oriented coordinate system, therefore the principal streamline direction, and two independent normal directions must be determined. The term derived by Probst was for an ϵ-based RSM turbulence model and it has been adapted to the ω length scale equation of the SST model, referred to as PG model. This PG model extension was used as a preliminary variant of sensitizing the model to the effects of irrotational strains and hence pressure gradients and it uses the gradients of the mean flow in order to extract the effects of irrotational strains as follows:

$$S_\omega = -C_\omega^* \frac{\omega}{k} \rho \left(\overline{u_s'^2} \frac{\partial U_s}{\partial x_s} + \overline{u_{n1}'^2} \frac{\partial U_{n1}}{\partial x_{n1}} + \overline{u_{n2}'^2} \frac{\partial U_{n2}}{\partial x_{n2}} \right) \tag{10}$$

The term S_ω added to the Eq. 2. The constant C_ω^* has a value of 0.58. Though this PG model extension led to significant improvements in several different test cases, it is not Galilean invariant due to its dependency on streamline oriented coordinate system. The property of a Galilean—invariant turbulence model is essential for general applications in aeronautical flows, e.g. in flow simulations with rotating aircraft components such as propulsive rotors, or for simulating the effect of overall aircraft rotation. An alternative term referred to as "Vorticity squared to Production of k difference" (V2Pkd) is therefore derived and it is also based on the work of Hanjalić and Launder, although devised in a cartesian frame of reference. The V2Pkd term is a subtraction of a strain rate-dependent scalar from a vorticity-dependent scalar, and a coefficient is used to regulate the term's intensity:

$$V2Pkd = C_{\omega 4} \rho (W_{ij} W_{ij} - S_{ij} S_{ij})$$
$$W_{ij} = \frac{1}{2}\left(\frac{\partial U_i}{\partial x_j} - \frac{\partial U_j}{\partial x_i}\right) \qquad S_{ij} = \frac{1}{2}\left(\frac{\partial U_i}{\partial x_j} + \frac{\partial U_j}{\partial x_i}\right) \tag{11}$$

W_{ij} is the mean vorticity tensor and S_{ij} is the mean strain rate tensor. $C_{\omega 4}$ is the term's coefficient. In equilibrium boundary layers, the term is essentially zero since W_{ij} and S_{ij} have approximately the same magnitude. For non-equilibrium flows, the predominance of its vorticity- and strain rate-components alternate and the sign of the term varies throughout the flow. When the V2Pkd term assumes a positive sign, it is interpreted as a source in the ω-equation 2 leading to an increase in dissipation and subsequent decrease in turbulent stresses. In contrast, when the term is negative, it acts as a sink leading to the opposite effect, i.e. decrease in dissipation and increase in turbulent stresses. It is of high importance to study and understand this term further using a variety of practical flows, to gain further insight.[1]

3 Calibration Using Bayesian Update

Three extensions are added to the standard Menter—SST model and in total five different coefficients need to be calibrated. It has to be noted that the extensions are applied simultaneously and treated as one whole extended model with additional five coefficients. The uncertainties present in the experimental data which are used for calibration are also included in the calibration process. First and foremost is the identification of the parameter space. During the development of the extensions, bounds of the respective coefficients are laid out based on literature and sensitivity studies (Table 1). For this parameter estimation procedure, a system with states $\mathbf{x} \in \mathbb{R}^d$ that depend on the uncertain parameters $\mathbf{q} \in \mathbb{R}^N$ is considered. In the present work, the states are the flow solver's (DLR-TAU) solutions, depending on the $N = 5$ parameters of the extended turbulence model as listed in Table 1.

$$\mathbf{x} : \mathbb{R}^N \longrightarrow \mathbb{R}^d \tag{12}$$

$$\mathbf{q} \mapsto \mathbf{x}(\mathbf{q}). \tag{13}$$

The *measurable* $\mathbf{y}(\mathbf{x}) \in \mathbb{R}^M$ is a quantity that depends on the system's states, and in CFD context the measurables are the flow field responses such as velocities, Reynolds stresses and surface pressure measured on a spatial grid. A measurement $\mathbf{y}_m \in \mathbb{R}^M$ is an instance of the measurable that was actually measured, and is usually modeled as the sum of the true measurable and an instance of an additive mean-free noise with distribution π_ϵ. Wind tunnel measurements provide the $\mathbf{y}_m$. Of course the choice of the measurables and their position are geometry-dependent. Bayes' formula expressed with $P(X)$ being the probability of the event X reads

$$P(X|Y) = \frac{P(Y|X)P(X)}{P(Y)} \tag{14}$$

[1] All the three model extensions were tested using a zero pressure gradient flat plate test case and have shown no adverse effects for attached turbulent shear flows in equilibrium.

Table 1 Initial intervals of model coefficients

Extensions	V2Pkd	SARC	SARC	SARC	SAS
Coefficients	$C_{\omega 4}$	c_{r1}	c_{r2}	c_{r3}	ζ_2
Left boundary	−0.5	0	0	0	0
Right boundary	10	1.5	10	1.5	7

and gives the probability of X if Y is given. Expressed in probability densities, the posterior distribution $\pi_{\overline{\mathbf{q}}} := \pi_{\mathbf{q}|\mathbf{y}_m}$ of the parameter $\mathbf{q}$ conditioned on our measured response $\mathbf{y}_m$ is given by Bayes' theorem as

$$\pi_{\overline{\mathbf{q}}} := \pi_{\mathbf{q}|\mathbf{y}_m}(\mathbf{q}) = \frac{\pi_{\mathbf{y}_m|\mathbf{q}}(\mathbf{q})\pi_{\mathbf{q}}(\mathbf{q})}{\int \pi_{\mathbf{y}_m|\mathbf{q}}(\mathbf{q})\pi_{\mathbf{q}}(\mathbf{q})\, d\mathbf{q}}. \tag{15}$$

Here $\pi_{\mathbf{q}}$ is the prior distribution of the parameter, e.g. an initial guess of uniform distribution in-between the bounds given in Table 1, based on professional expertise and $\pi_{\mathbf{y}_m|\mathbf{q}}$ is the likelihood, the probability of measuring $\mathbf{y}_m$ given a certain value of the parameters $\mathbf{q}$. The integral in the denominator is the multidimensional integral taken over all components and values of $\mathbf{q}$: $\int \pi_{\mathbf{y}_m|\mathbf{q}}(\mathbf{q})\pi_{\mathbf{q}}(\mathbf{q})d\,\mathbf{q} = \int_{\infty} \ldots \int_{\infty} \pi_{\mathbf{y}_m|\mathbf{q}}(\mathbf{q})\pi_{\mathbf{q}}(\mathbf{q})\, dq_1 dq_2 \ldots dq_M$, where q_i denotes i-th component of the vector $\mathbf{q}$. The likelihood is expressed from the distribution of the measurement noise,

$$\pi_{\mathbf{y}_m|\mathbf{q}}(\mathbf{q}) = \pi_{\epsilon}(\mathbf{y}_m - \mathbf{y}(\mathbf{q})), \tag{16}$$

which is assumed to be mean-free. A one point estimate of the parameter can be given by the most probable value given the measurement

$$\mathbf{q}_{\mathrm{MAP}} = \arg\max_{\mathbf{q}} \pi_{\mathbf{q}|\mathbf{y}_m}(\mathbf{q}), \tag{17}$$

the so-called maximum a-posteriori (MAP) estimate.

In general, the calibration process is performed as follows. Using CFD simulations, sensitivities studies are performed for individual extensions to determine the lower and upper bound of each coefficient. Sample sets of coefficients are generated by quasi Monte Carlo sampling within this parametric space and CFD simulations using DLR—TAU code are performed for these sample points using the extended model. For the case of Backward Facing Step, 300 CFD simulations were performed as an example. Based on this data, a surrogate model is built for the given test case, or if required for a combination of test cases using corresponding datasets. During the process of surrogate modeling different approaches were tested and finally generalized Polynomial Chaos (gPC) was chosen as a suitable fit. CFD data set of 300 simulations are used to train and validate the surrogate model and this surrogate model replaces the costly CFD test runs. Finally the posterior distribution is obtained

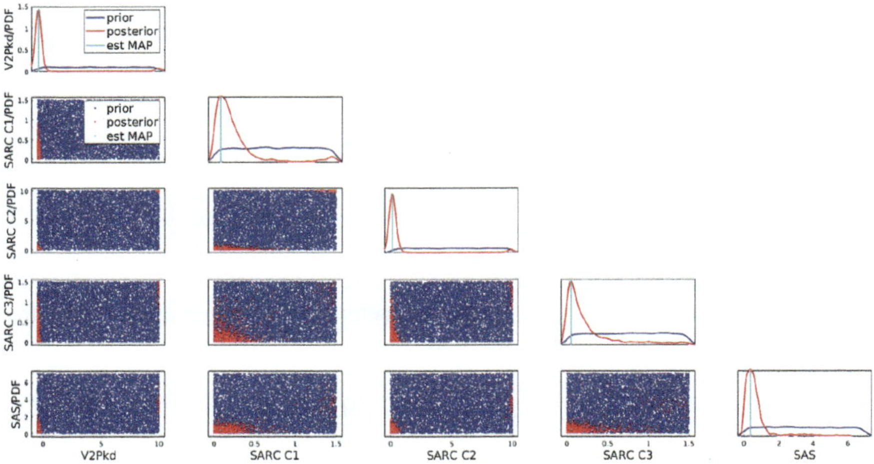

Fig. 1 Combined Bayesian update of Model53 delta wing at angle of attack 20°, 25° and 28°

Table 2 Calibrated parameters of the turbulence model extensions

Extensions	V2Pkd	SARC	SARC	SARC	SAS
Coefficients	$C_{\omega 4}$	c_{r1}	c_{r2}	c_{r3}	ζ_2
Set 1	1.82	1.46	3.29	0.01	2.18
Set 2	−0.42	0.08	0.15	0.05	0.32
Set 3	−0.43	0.09	0.14	0.05	0.46

as shown in the Fig. 1 and the maximum a-posteriori coefficient set is obtained, which would serve as the coefficients of the extended turbulence model. Table 2 provides calibrated coefficients for different test cases. Successful calibration was performed using individual test geometries under different flow spectrum [12]. Set 1 provides the data for Backward Facing Step, transonic axisymmetric bump and Model 53 delta wing at 20° angle of attack. Set 2 is the date for Model 53 delta wing at 25° and 28° angles of attack, near to the vortex breakdown region. Set 3 is the calibration of Model 53 at 20°, 25° and 28° angles of attack. Hence set 3 covers a rather large variation of vortex break down along the lift curve, to enable prediction of that flow phenomenon.

4 Model 53 and Model 56 Delta Wing

Two types of delta wings are analyzed, associated with complex vortical flows. The leading edge profile is rounded and therefore the position of the onset of vortex flow is not fixed at the apex. Moreover the presence of a leading edge slat presents complex vortex interaction scenario, which makes this an interesting and challenging geometry

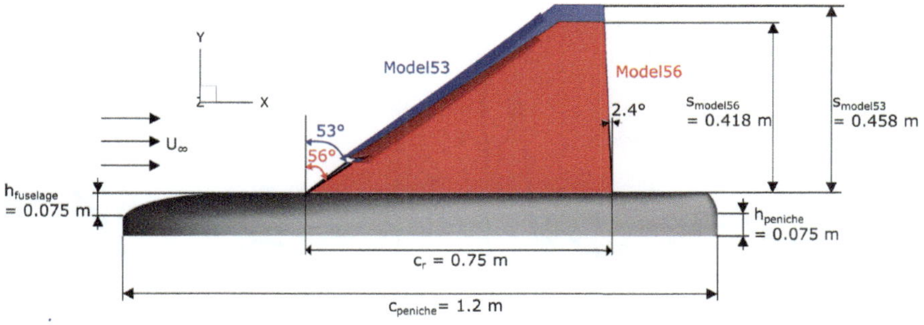

Fig. 2 Model53 and Model56 W/T model sketch, top view xy plane [1]

to pursue. The Model53 (Fig. 2) is a wind tunnel model representing a generic delta wing configuration with a 53° sweep angle, a tip twist of 4° and a slat deployed at 20°. A sweep angle variation of the same configuration is named Model56 (Fig. 2), with a 56° sweep angle. Additive manufacturing technology was used to produce the wind tunnel models [13]. The experimental analyses of the wind tunnel models were made available from Technical University of Munich, measured in the Wind Tunnel A (W/T-A) facility of the Chair of Fluid Mechanics. The freestream velocity is slightly adjusted in order to match the Mach number of 0.15, which gives a velocity of 52m/s. A Reynolds number of $1.7 \cdot 10^6$ based on l_μ is thereby reached.

Numerical simulations were performed using the DLR—TAU flow solver. Unsteady RANS simulations were performed with a numerical mesh of around 8.85 million points, using low Mach number preconditioning. Second order Central scheme was used for spatial discretization and backward Euler was used for temporal discretization. The average y^+ value of the numerical grid over the wing surface was about 0.5 for 20° α case. Sensitivities with respect to the used numerical grid was already studied [14] and the calibration process using different mesh levels was also validated [12].

4.1 Model 53 Calibration

A detailed calibration study was performed initially using set 1 and set 2 and valuable insight could be gained [12]. The target of the work was to calibrate the model with Model 53 delta wing under different vortical flow conditions. 110 simulations were therefore performed for each of 20°, 25° and 28° angle of attack. The generated data set was used in the Uncertainty Quantification (UQ) suite to build the surrogate model and calibrate the model to find the MAP point, which are the coefficients of set 3 given in Table 2. Spanwise surface pressure distribution coefficient is shown in Fig. 3 and it can be seen that the standard Menter—SST model in itself has a good agreement with the experimental data. Both set 2 and set 3 calibrated coefficients

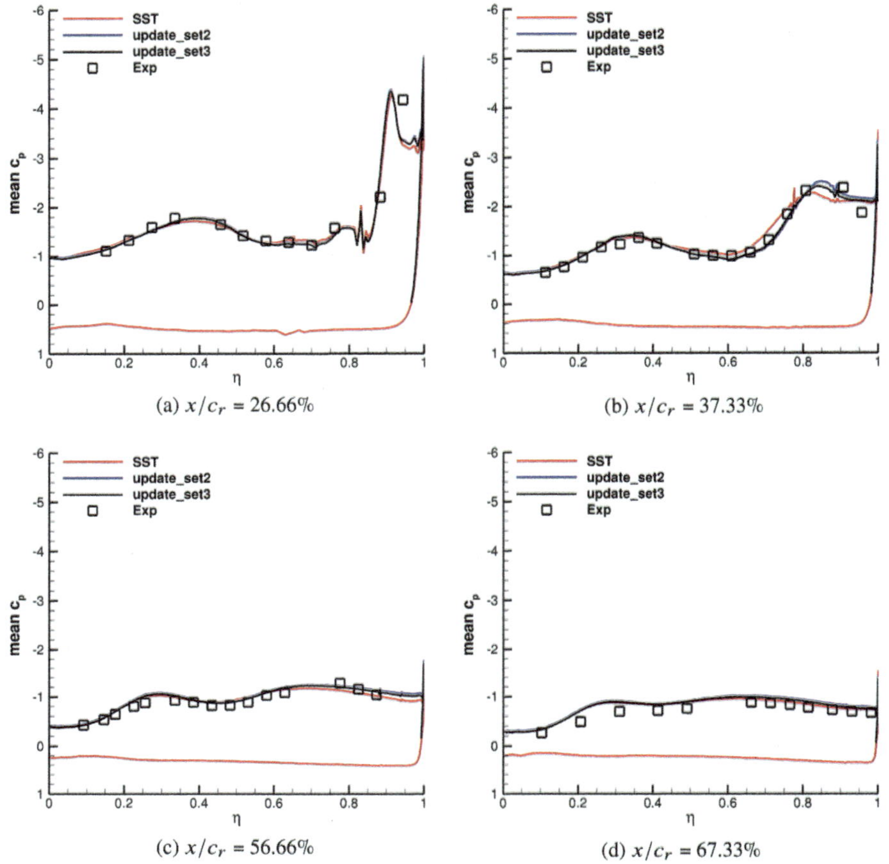

Fig. 3 Comparison of original and calibrated turbulence model for Model53 delta wing at $\alpha = 25°$

have good agreements with measurements throughout the delta wing surface. Around 30 to 40% of the root chord length, the characteristics of the slat vortex are improved by both the calibrated coefficients near to the wing cut section tip.

Iso surfaces of vortices using non dimensional Q-criterion at 20° angle of attack, flooded transparently by the non dimensional pressure coefficient are shown in Fig. 4. Surface skin friction streamlines are superimposed on the upper surface of the wing, illustrating the standard Menter—SST model and extended model calibrated with coefficients of set 3. Increase in suction peak of the apex vortex for the extended model can be seen in Fig. 4b along with a minor increase in the thickness of the vortex sheet. Footprint of the apex vortex is slighty shorter for the extended model in comparison to the standard model. Apart from a couple of straight running streamlines near the wing root, the dividing streamlines are deflected towards the core of the apex vortex caused by the low pressure underneath the vortex. Suction peak in the onset region of the slat vortex appears to be same between the models and slat vortex of the calibrated

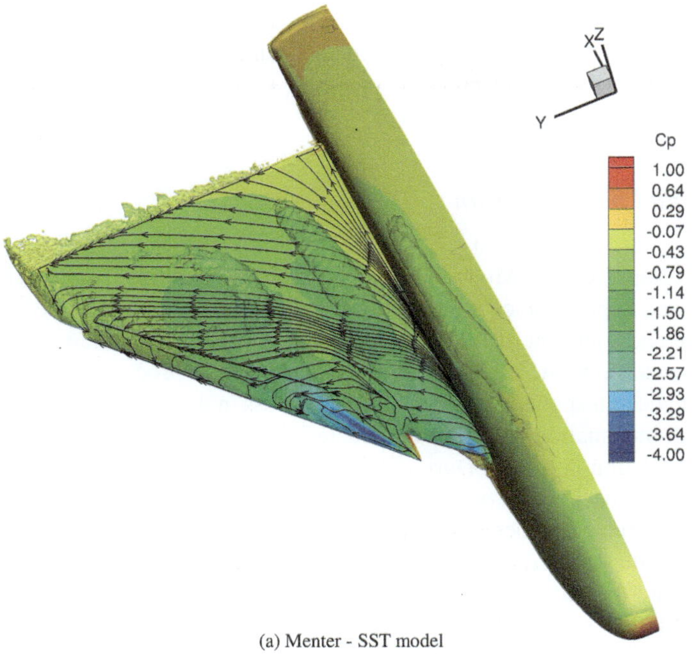

(a) Menter - SST model

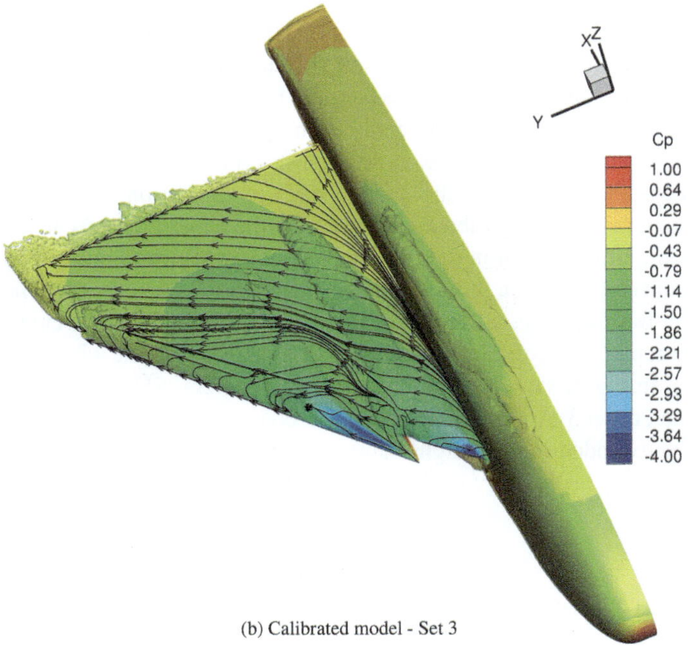

(b) Calibrated model - Set 3

Fig. 4 Isosurfaces using nondimensional Q criterion (0.0025) at $\alpha = 20°$, along with surface streamlines over the Model53 wing model appears to be slightly thicker. Vortex breakdown is observed around the mid wing section and onset of vortex location and shape of vortex breakdown appear to be very similar for both the models.

4.2 Model 56 Validation

Model56 is derived from Model53 by varying the sweep angle. Vortex behavior and its associated dependencies are largely influenced by the sweep angle, in terms of magnitude and footprint of the vortices. Purpose of using Model56 as a validation case is to determine the effectiveness of the calibration process and further use in subsequent design applications. Calibration process in the UQ suite does not include any a priori information on the flow characteristics of Model56 and therefore it suits well for validation purposes. During the VitAM project phase, an alternative model calibration approach based on deterministic optimization of model coefficients was studied by Airbus Defence and Space and Technical University of Munich. The results from the standard Spalart Allmaras (SA) turbulence model and an optimized SA-model are therefore added as a reference [1]. Test conditions remain the same as that of Model53. Unsteady RANS simulations were performed with a numerical mesh of 10.21 million points, with same numerical settings as that of Model53 test case. The average y^+ value of the numerical grid over the wing surface was about 0.7 and about 1 in the vortex onset regions for 25° α case.

Spanwise surface pressure distribution coefficient is shown in Fig. 5 and it can be seen that the standard Menter—SST model has a good agreement with the measurements, similar to the Model53 calibration case. Around 30% of the root chord length, improvements in both apex vortex and slat vortex suction peak could be seen by the calibrated models for both set 2 and set 3. In this region, the footprint of both vortices seem to be slightly inboard compared to the experiment. At 40% of the root chord length, improvement in the suction peak of slat vortex is more apparent. This improvement could be seen further downstream around 53% of the root chord length. Near to the trailing edge, all the simulated results are in good agreement showing the dissipation of vortices. Surface pressure distribution over the wing surface is also reproduced in Fig. 6, showing both the standard Menter—SST and extended model with coefficient set 3. Apart from the pronounced increase in slat vortex suction pressure for the extended model, slight increase in both apex and slat vortex footprints could be seen.

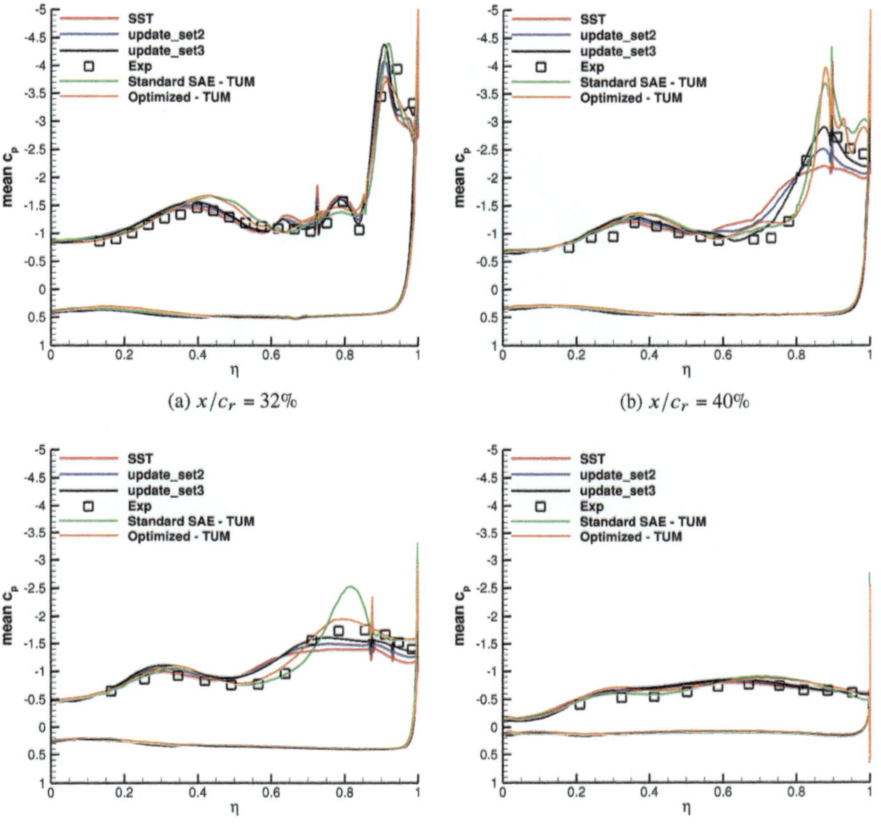

(a) $x/c_r = 32\%$

(b) $x/c_r = 40\%$

(c) $x/c_r = 53.33\%$

(d) $x/c_r = 73.33\%$

Fig. 5 Comparison of original and calibrated turbulence model for Model56 delta wing at $\alpha = 25°$

5 Summary and Outlook

Flow physics based modeling extensions are explored and implemented to standard Menter—SST turbulence model, to improve the prediction capabilities at challenging flow conditions. Bayesian method based on Uncertainty Quantification is used to calibrate the extended model based on data sets from complex delta wing simulations under vortical flow conditions. Improvements in the numerical solution are illustrated along with validation of the calibration procedure, serving as a benchmark procedure for further applications. More insight could be gained by analyzing force and moment coefficients on pitch break location and the associated vortex breakdown dynamics. An important aspect of the calibration process is not only the maximum a—posteriori (MAP) point, but the whole posterior distribution of the calibrated coefficients. Data points of coefficients neighboring the MAP point could potentially offer better numerical solutions, and this should be explored. Development of

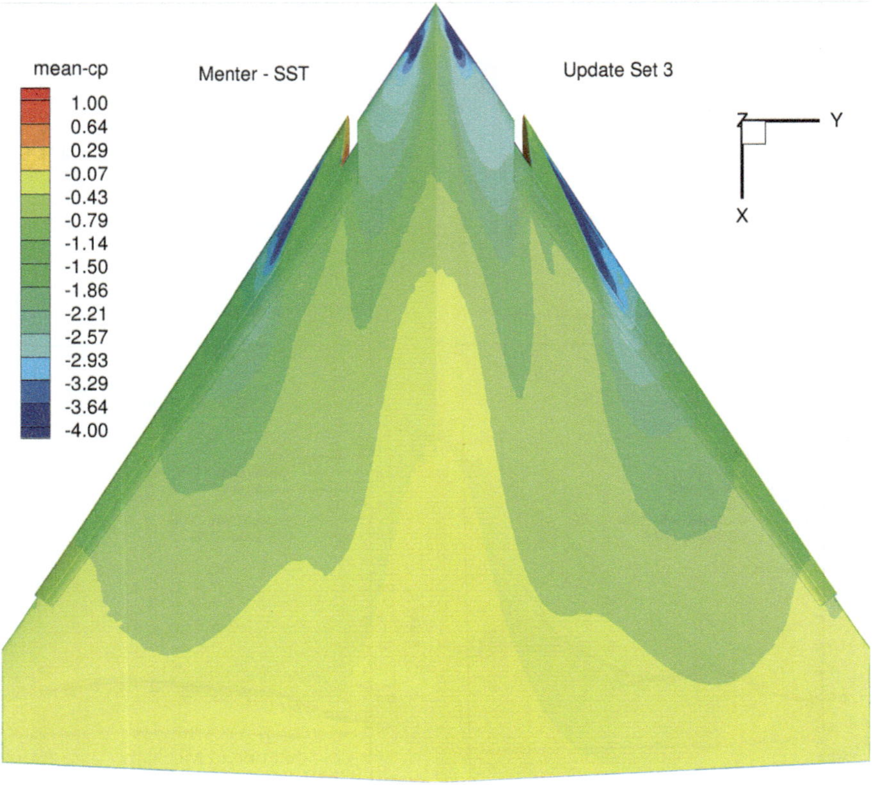

Fig. 6 Surface pressure distribution of Model56 at $\alpha = 25°$

an automated calibration system in which the coefficients of interest in the posterior distribution could be put to test initially using surrogate model and validated using CFD simulation, is also envisioned.

Acknowledgements The authors gratefully acknowledge LuFo V—"Vitual Aircraft Model-Calibration" (LUFOV2-790-024/FKZ:20A1504A) for funding this research project. The turbulence modeling group of the VitAM project is acknowledged for the numerous discussions, knowledge and data transfer. DLR is acknowledged for the TAU code and informative discussions. NASA turbulence resource homepage is acknowledged for its numerical meshes and experimental data. The computations shown in this paper were performed using the HLRN (Norddeutscher Verbund für Hoch-und Höchstleistungsrechnen), which was essential for the cases requiring large resources. Development of the here used UQ suite was supported through "2018-1.2.1-NKP-00008 Exploring the Mathematical Foundations of Artificial Intelligence".

References

1. Moioli, M., Breitsamter, C., Sørensen, K.: Turbulence modeling for leading-edge vortices: an enhancement based on experimental data. AIAA J. (2022). https://doi.org/10.2514/1.J060062
2. Knight, D., Saffman, P.: Turbulence model predictions for flows with significant mean streamline curvature. In: 16th Aerospace Sciences Meeting (1978). https://doi.org/10.2514/6.1978-258
3. Spalart, P.R., Shur, M.: On the sensitization of turbulence models to rotation and curvature. Aerosp. Sci. Technol. (1997). https://doi.org/10.1016/S1270-9638(97)90051-1
4. Smirnov, P.E., Menter, F.R.: Sensitization of the SST turbulence model to rotation and curvature by applying the spalart? Shur Correction Term. J Turbomach. (2009). https://doi.org/10.1115/1.3070573
5. Shur, Michael L., Strelets, M.K., Travin, A.K., Spalart, P.R.: Turbulence modeling in rotating and curved channels: assessing the spalart-shur correction. AIAA J. (2000). https://doi.org/10.2514/2.1058
6. Egorov, Y., Menter, F.: Development and application of SST-SAS turbulence model in the DESIDER project. Adv. Hybrid RANS-LES Model. (2008). https://doi.org/10.1007/978-3-540-77815-8_27
7. Jakirlić, S., Maduta, R.: Steady RANS modeling for improved prediction of wall-bounded separation. AIAA J. (2016). https://doi.org/10.2514/1.J054399
8. Cécora, R.-D., Reynolds-Spannungsmodell der Turbulenz für Anwendungen der Flugzeugaerodinamik, Dissertation—Technische Universität Braunschweig (2015). ISBN 978-3928628785
9. Probst, A., Reynoldsspannungsmodellierung für das Überziehen in der Flugzeugaerodynamik, Dissertation—Technische Universität Braunschweig (2013). ISBN 978-3844021233
10. Hanjalić, K., Launder, B.E.: Sensitizing the dissipation equation to irrotational strains. J. Fluids Eng. (1980). https://doi.org/10.1115/1.3240621
11. Subbian, G., Radespiel, R.: Assessment of Extensions for an Eddy Viscosity Turbulence Model for Vortical Flows, New Results in Numerical and Experimental Fluid Mechanics XII (2020). ISBN=978-3-030-25253-3
12. Subbian, G., Botelho e Souza, A. C., Radespiel, R., Zander, E., Moshagen, T., Friedman, N., Moioli, M., Breitsamter, C., Sørensen, K.: Calibration of an extended eddy viscosity turbulence model using bayesian update. AIAA J. (2022). https://doi.org/10.2514/1.J060527
13. Moioli, M., Reinbold, C., Sørensen, K., Breitsamter, C.: Investigation of additively manufactured wind tunnel models with integrated pressure taps for vortex flow analysis. Aerosp. J. (2019). https://doi.org/10.3390/aerospace6100113
14. Moioli, M., Breitsamter, C., Sørensen, K.: Parametric data-based turbulence modelling for vortex dominated flows. Int. J. Comput. Fluid Dyn. (2019). https://doi.org/10.1080/10618562.2019.1617857
15. Menter, F.: Zonal two equation k-ω turbulence models for aerodynamic flows. In: 23rd Fluid Dynamics, Plasmadynamics, and Lasers Conference (1993). https://doi.org/10.2514/6.1993-2906

VitAM-Flex—Computational Modelling and Simulation to Study Effects of Elastic Aircraft Structures on Flight Physics

Andreas Hermanutz and Mirko Hornung

Abstract Different computational approaches to predict aircraft loads and aeroelastic responses based on industry-relevant demonstration examples are presented. To reduce computational efforts, methods to establish a structural reduced-order model and its applicability for flexible aircraft configurations are studied. To gain a better understanding of the underlying flight physics, higher-order aeroelastic modelling approaches are used through coupled CFD-CSM techniques. Lastly, the impact of nonlinearities owing to large deformations on the prediction accuracy when using state-of-the-art flutter prediction methods is analysed.

Keywords Aircraft structural design · Aeroelasticity · Computational flutter analysis · Reduced order model

1 Introduction and Motivation

Increasingly stringent requirements such as low fuel consumption and low emissions at high ranges and transport capacities require precise predictions of the aircraft characteristics early in the development process. The necessary predictions for design can be obtained from both experiments and simulations. However, as experiments are usually associated with high costs, simulations play an increasingly important role. Notably, the possibility of accurately and comprehensively describing an aircraft and its characteristics in the form of a virtual aircraft model (VAM) plays a key role in aircraft development. If the interdependence of all relevant disciplines is represented in this model with suitable accuracy at an early stage, virtual flight tests can be realised prior to the first flight.

A. Hermanutz · M. Hornung (✉)
Chair of Aircraft Design, Boltzmannstr. 15, 85748 Garching b. München, Germany
e-mail: Mirko.Hornung@tum.de

A. Hermanutz
e-mail: Andreas.Hermanutz@tum.de

R. Heinrich (ed.), *Advanced Aircraft Understanding via the Virtual Aircraft Model*, Notes on Numerical Fluid Mechanics and Multidisciplinary Design 155,
https://doi.org/10.1007/978-3-031-69425-7_10

Depending on the research objective, different models with different requirements in terms of the accuracy and simulation time are required for the development process. Representing the investigation of flight-physical phenomena in an unified model is difficult when using existing design and simulation methods. Thus, in VitAM-Flex, tasks to solve high-integral aeroelastic design methods using reduced-order models (ROM) as well as methods from strongly coupled CFD-CSM simulations are investigated. The studied computational models and simulation tools are used to gain a deeper understanding of the underlying flight physics. Furthermore, the idea of a VAM is investigated and demonstrated through on industrially significant use cases.

The need for novel aircraft configurations with increasingly lower structural weights and highly efficient aerodynamic designs is growing; resulting in more flexible wings. The associated aeroelastic effects have an influence over the aerodynamic efficiency, control surface effectiveness, and structural loads. Aeroelastic stability criteria such as wing flutter and divergence have also been emphasised. Common technical procedures are applied during the design process; these can be deemed as the industry standards at the time of the project. These procedures include linear methods from structural and fluid mechanics, such as the linear *finite element method* (FEM), *vortex lattice method* (VLM), and *doublet lattice method* (DLM). From a scientific perspective, approaches that are methodologically based on our partners' research projects, AeroGust [1] and the DLR-led project VicToria [2], were implemented. This includes methods from model reduction and high-fidelity simulations employing strongly coupled CSM-CFD.

2 Computational Modelling Setup

A generic reference aircraft configuration is used as a engineering example, with the CAD model serving as the starting point for model setup. As shown in Fig. 1, a generic high-aspect-ratio transport aircraft (TA) configuration is used. The corresponding wing has a size of $S = 58[\text{m}^2]$, $AR = 14.0$, and $TR = 0.2$, with a sweep of $\Lambda = 25°$.

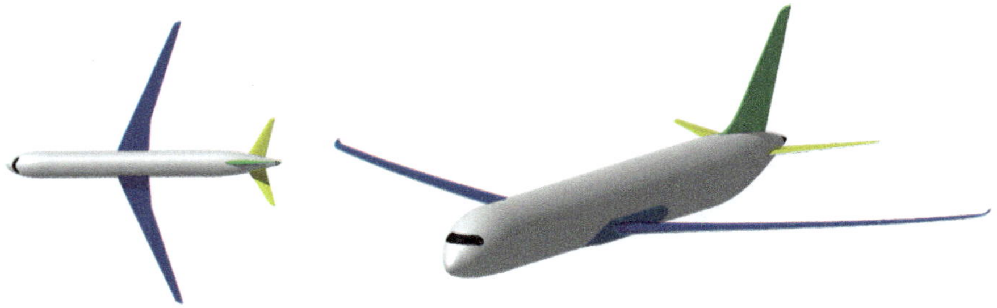

Fig. 1 Generic transport aircraft demonstrator configuration

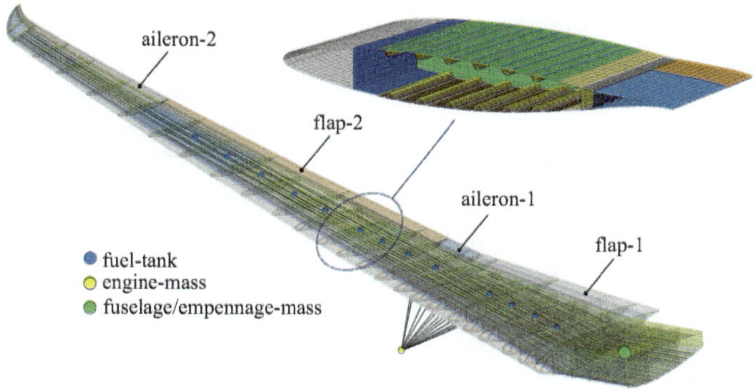

Fig. 2 TA aircraft wing FEM model

2.1 *Computational Structural Modelling Setup*

Composite materials are increasingly being used in aircraft, and the structural modelling of large components is performed on a macroscopic level by discretising the components with linear shell elements. The classical laminate theory is used to describe the anisotropic stiffness of the laminate definition. Within VitAM-Flex, the interactions between flexible structures and aerodynamics are primarily related to the wing structure. Therefore, the fuselage and empennage are modelled as rigid-body components to perform rigid-body movements and trim calculations.

Figure 2 shows the detailed FE model used for the structural simulation of the TA configuration. Because high-fidelity simulations are primarily conducted, only symmetrical flight manoeuvres are considered to reduce the required computational resources. Thus, with the corresponding symmetrical boundary condition, a half-model consists of 662040 degrees of freedom (DoFs). In terms of the discretisation with shell elements, offsets from the element reference plane were not considered in the modelling approach. Important joint areas such as the skin to spar flange or stringer to the skin are considered by a shared definition of the stiffness proportions. The overall global stiffness characteristic is represented correctly, whereby discontinuities owing to jumps in stiffness are considered more closely in the structural dimensioning analysis and corrected if necessary.

2.2 *Reduced-Order Modelling*

The VitAM-Flex project is investigating how reduced order structural models can be effectively used to describe the physics of flight using a VAM. Currently, aeroelastic simulations encounter the problem that thousands of simulations need to be performed to describe the entire flight envelope, including all flight loads and cargo configurations. The available computational resources are not sufficient for this purpose. Therefore, faster simulation methods are still being developed and explored. In particular, for simulations in the time domain, such as those involving gusts or manoeuvres, including flight control systems, reduced models are indispensable. Although the highest accuracy is at the foreground for higher-fidelity methods, computational costs are equally important for reduced-order methods. This leads to a conflict between the objectives, as the methods must both be computationally fast and not compromise on relevant accuracy.

It is inevitable that, when a highly detailed model is reduced, information is simultaneously lost. Therefore, the challenge lies in using reduced methods that meet the requirements for computational performance and investigate the physical effects with sufficient accuracy. For structural modelling, condensation methods are studied to reduce the corresponding systems of equations to be solved and, consequently, the degrees of freedom of the system. Accordingly, known algorithms are implemented and examined using a software tool to investigate their applicability in aircraft load simulations. The following condensation methods are considered [3–5]:

- Guyan reduction (GRM)
- Dynamic reduction (DRM)
- Improved reduced system (IRS).

The system to be solved is usually reduced in terms of its dimension to a few hundred DoFs, which are then grouped in the so-called analysis set; the neglected DoFs are grouped in the omitted set. To determine the reduced analysis set, the linear transformation matrix [T] from the full set to the reduced DoF set is calculated as followed:

$$\begin{pmatrix} \mathbf{x}_a \\ \mathbf{x}_o \end{pmatrix} = \begin{bmatrix} T \end{bmatrix} \mathbf{x}_a. \tag{1}$$

In this context, the indices a and o represent the **analysis** and **omitted** sets, respectively.

Guyan reduction method (GRM): Guyan reduction is one of the most commonly used reduction methods. Considering only the stiffness matrix, and thus neglecting dynamic effects, this is referred to as the static condensation method. The static system is separated into individual DoF sets, such that the static structural equilibrium equation can be rewritten as follows:

$$\begin{bmatrix} K_{aa} & K_{ao} \\ K_{oa} & K_{oo} \end{bmatrix} \begin{pmatrix} \mathbf{x}_a \\ \mathbf{x}_o \end{pmatrix} = \begin{pmatrix} \mathbf{F}_a \\ \mathbf{F}_o \end{pmatrix} \tag{2}$$

Solving Eq. 2 with respect to the analysis set yields the following transformation matrix [T]:

$$\begin{pmatrix} \mathbf{x}_a \\ \mathbf{x}_o \end{pmatrix} = \begin{bmatrix} I \\ -K_{oo}^{-1} K_{oa} \end{bmatrix} \mathbf{x}_a. \tag{3}$$

The transformation of the mass matrix is performed analogously. The static results are represented exactly as the GRM uses as input. However, as described [6], errors occur owing to an incorrect representation of the reduced mass matrix, which affects the dynamic response characteristics in particular.

Dynamic model reduction (DRM): Dynamic model reduction is based on the static Guyan reduction, except that the mass matrix is included in the computation of the transformation matrix. The equation of motion is transferred to the modal space, and the transformation matrix is expressed as follows:

$$\begin{pmatrix} \mathbf{x}_a \\ \mathbf{x}_o \end{pmatrix} = \begin{bmatrix} I \\ -(K_{oo} - \omega^2 M_{oo})^{-1}(K_{oa} - \omega^2 M_{oa}) \end{bmatrix} \mathbf{x}_a . \tag{4}$$

Equation 4 indicates that the influence of the mass matrix is considered relative to ω. In the case of $\omega = 0[Hz]$, DRM leads to the same result as GRM. The major disadvantage of this method, however, is that the transformation matrix is designed only for a certain ω. For applications where multiple eigenmodes in a larger range must be included to correctly describe the structural motion, DRM is prone to errors when obtaining a general solution.

Improved reduced system (IRS): IRS employs the output of the static GRM model and adds an additional correction matrix to account for dynamic effects. The transformation matrix can be expressed as follows:

$$\left[T_{IRS}\right] = \left[T_{GRM}\right] + \left[T_{corr}\right]$$

$$\left[T_{IRS}\right] = \begin{bmatrix} I \\ -K_{oo}^{-1} K_{oa} \end{bmatrix} + \begin{bmatrix} 0 \\ \left[K_{oo}^{-1} M_{oa} - K_{oo}^{-1} M_{oo} K_{oo}^{-1} K_{oa}\right]\left[D\right] \end{bmatrix} \tag{5}$$

An additional matrix $[D]$ can be used to eliminate the frequency dependency of the DRM [5].

2.3 Computational Aerodynamic Modelling Setup

The aerodynamic modelling used for VitAM-Flex can be separated into two categories. These methods are based on the potential theory, such as VLM and DLM, and the solution of Reynolds-averaged Navier-Stokes (RANS) equations using computational fluid dynamics (CFD). For low-speed conditions, the corresponding methodological physical modelling assumptions of the potential theory, such as the compressibility at a low Mach numbers are sufficiently reflected. However, flow detachment phenomenon is not considered. Therefore, high-fidelity CFD methods are applied.

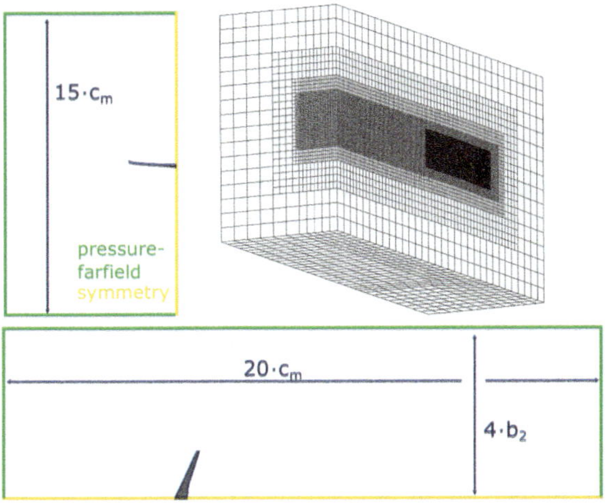

Fig. 3 TA aircraft wing aerodynamic CFD model

The discretisation of the flow domain used for aerodynamic modelling based on the RANS methods is performed on an unstructured Cartesian grid, as shown in Fig. 3. The finite volume (FV) solver implemented in FLUENT is used to solve the constitutive laws. The discretisation comprises $\approx 12 \cdot 10^6$ cells. This represents a compromise between computational accuracy and computational costs. The near-wall boundary layer flow is not completely resolved because the investigations focus on structurally effective normal forces, that is, those resulting from the pressure load. Turbulence is modelled with the *SST* model combined with an additional wall function correction.

2.4 Computational Aeroelastic Modelling Setup

High-fidelity methods are primarily applied to quasi-steady-state flight manoeuvres, with the aim of studying critical quasi-steady sizing loads and accurate prediction methods for wing structural deformations. In particular, the interaction with aerodynamics is considered for highly elastic wing configurations, in order to predict the wing's flight shape under different manoeuvre load cases. For this purpose, a coupled CFD-CSM simulation is used. An implicit, strongly coupled sequential solution algorithm was implemented in the C++-based computational framework, with the basic procedure as shown schematically in Fig. 4 [8].

The transfer of interface forces $\mathbf{F}$ and displacements $\mathbf{u}$ is performed by mapping different grids based on the shape functions. This mathematical mapping, where $[A]$ is the mapping matrix, can be expressed as in Eq. 6 [9].

$$\mathbf{u}_{CFD} = [A]\,\mathbf{u}_{CSM} \qquad \mathbf{F}_{CSM} = [A]^T\,\mathbf{F}_{CFD} \tag{6}$$

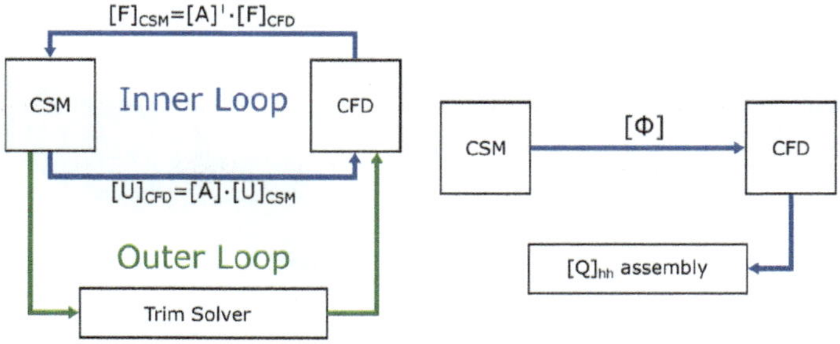

Fig. 4 Fluid interaction coupling scheme (left), structural eigenmode mapping scheme (right)

To investigate quasi-free-flying aircraft, the existing coupling algorithm is supplemented with an additional trim routine. For this purpose, the implicit solution process is separated into two sub-loops: the inner loop, in which the coupled CSM-CFD model is solved, and an additional outer loop, in which the trim variables are recalculated and updated. Because both loops are strongly coupled, which means that they interfere with each other, numerical stabilisation is applied trough under-relaxation of the interface variables. This is particularly necessary at the start of the simulation, when little information is known about the equilibrium state. The trim variables are updated in each iteration step until convergence in both the trim and flow fields is achieved.

3 Virtual Aircraft Model Using ROM Methods

The setup and verification of the structural ROM models are based on a comparison with the full structural TA model. For this purpose, it is assessed whether the mathematical description of the structural-mechanical system reflects the full structural model's response under static and dynamic conditions. In particular, for the case of model reduction, it is expected that partial information will be lost. To apply the methods with sufficient accuracy, the deviations of the relevant characteristics from those in the full TA model are compared.

First, the static deformation behaviour between the overall and reduced models is examined using the TA-Wing as an example. As shown in Fig. 5 (left), the reduction points along the wing span are simultaneously used as the load application points in the full model. The load is transferred locally to the wing box via RBE3 interpolation elements, as shown in Fig. 5 (right).

The aerodynamic forces are obtained according to the strip theory at $Ma = 0.3$ and a flight altitude of $h = 1000\,[\text{m}]$. The wing tip deformation is considered as the evaluation reference point, and the results for the different reduction methods are listed in Table 1. The static deformation response is well reproduced for all the

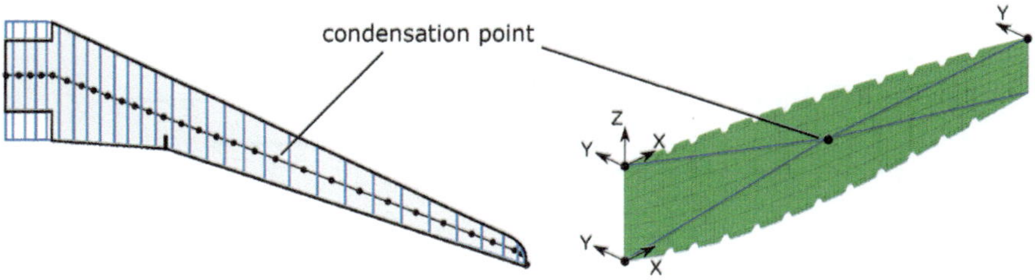

Fig. 5 Fluid interaction coupling scheme (left) and structural eigenmode mapping scheme (right)

Table 1 Comparison of static wing tip deformation

Displacment		Deviation		
DoF	Full FEM	GRM (%)	DRM (%)	IRS (%)
u_x	458.6 [mm]	0.0	0.0	0.0
u_y	−1272.4 [mm]	−0.00031	−0.00031	−0.00031
u_z	9246.6 [mm]	−0.00025	−0.00030	−0.00025
rot_x	0.664 [rad]	−0.00039	−0.00048	−0.00039
rot_y	−0.217 [rad]	−0.0006	−0.00078	−0.0006
rot_z	−0.054 [rad]	−0.00043	−0.000652	−0.00043

Table 2 Comparison of the natural frequency—deviation from the TA full model in %

Eigenmode	GRM (%)	DRM (%)	IRS (%)
1	0.0059	0.0003	0.0017
2	0.1035	0.0852	0.0002
3	0.0116	0.0106	0.0005
4	0.4334	0.4168	0.0001
5	2.3620	2.3324	0.0022
6	1.4476	1.4300	0.0011
7	1.7861	1.7801	0.0185
8	3.6551	3.6379	0.0319
9	8.5736	8.5500	0.1309
10	13.7807	13.7614	0.5728

three methods because only minor deviations in both the translational and rotational degrees are observed.

The dynamic responses are compared based on the eigenfrequencies and natural modes of the wing. The results for the first ten elastic natural frequencies of the wing are listed in Table 2. It can be noted that the IRS method represents the full model more closely than GRM and DRM. As the frequency increased, a deviation of up to ≈13% is observed in mode ten for the GRM and DRM.

Fig. 6 Eigenmodes MAC comparison TA versus Guyan reduction method

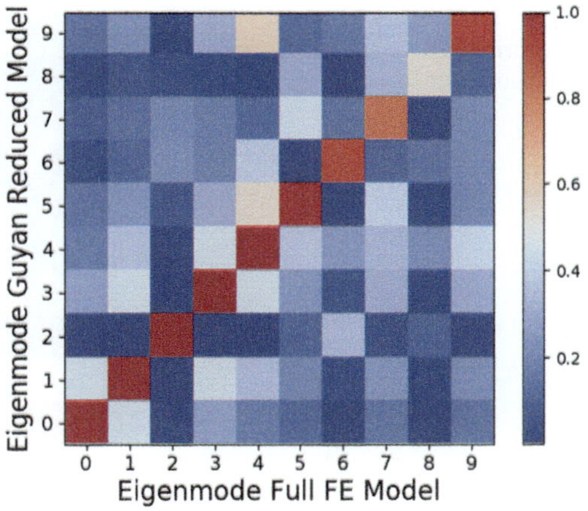

Fig. 7 Eigenmodes MAC comparison TA versus dynamic reduction method

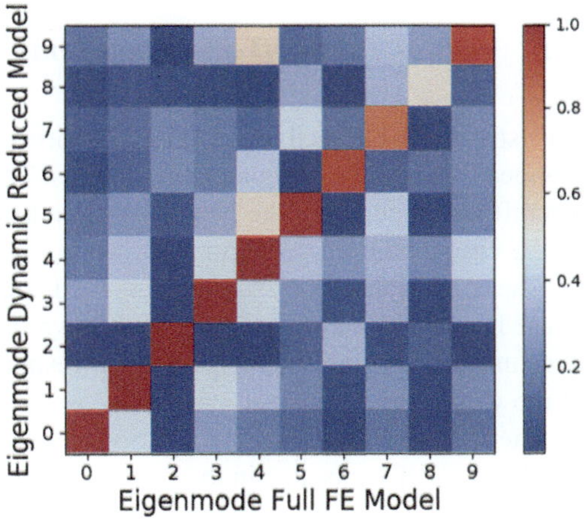

The aeroelastic response is dependent on the deformation, which can be characterised with the natural eigenmodes of the wing. To evaluate the similarity of the eigenmodes, the modal assurance criterion (MAC)

$$MAC = \frac{|\Phi_{full} \cdot \Phi_{red}|^2}{|\Phi_{full}| \cdot |\Phi_{red}|} \tag{7}$$

is used. The results are shown in Figs. 6, 7 and 8. Similar to the eigenfrequencies, the eigenmodes for the GRM and DRM deviate increasingly from the original full model. However, good agreement is noted for the case of the IRS method.

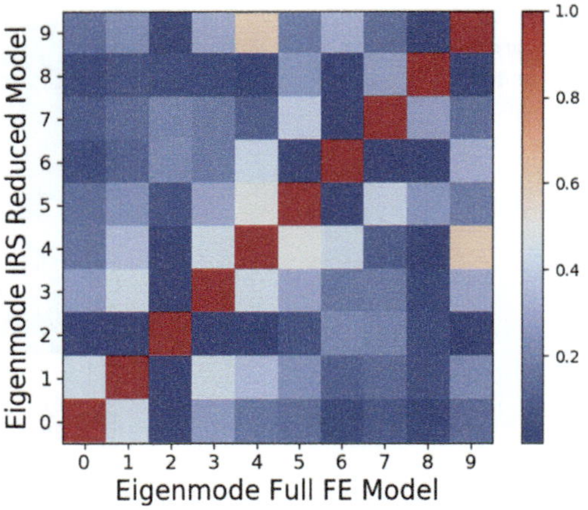

Fig. 8 Eigenmodes MAC comparison TA versus IRS reduction method

3.1 *Virtual Take-Off and Gust Load Simulation*

The structural reduced order methods are examined using a transient take-off manoeuvre and gust encounter simulations. The reduced equation of motion for the aircraft is solved using the Newmark-Beta time integration method. The sequence of the virtual take-off scenario is shown in Fig. 9.

The structural response of the wing tip displacement owing to manoeuvre loading is used as an evaluation example. As shown in Fig. 10, after the aircraft accelerated and finished rotating, the displacement response transitioned to a quasi-stationary climb with a short transient overshoot.

The second demonstration scenario involves a transient gust encounter. According to CS-25 [7], a 1-cos gust in the form of Eq. 8 is used for the transient gust profile.

$$v_g(t) = \frac{v_{g,0}}{2}\left(1 - cos\left(\frac{2\pi V}{l_g}t\right)\right) \tag{8}$$

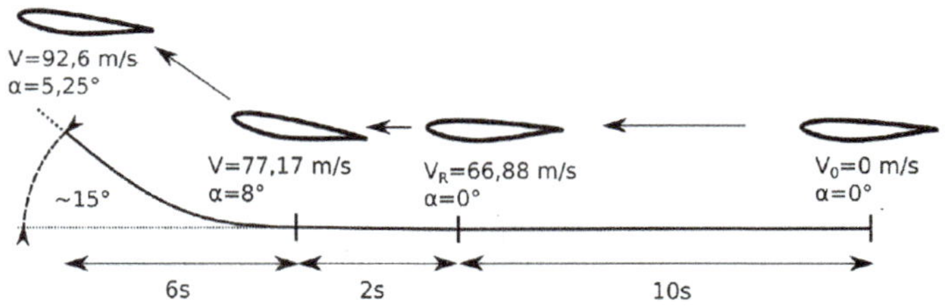

Fig. 9 Reduced Order model manoeuvrer flight scenario

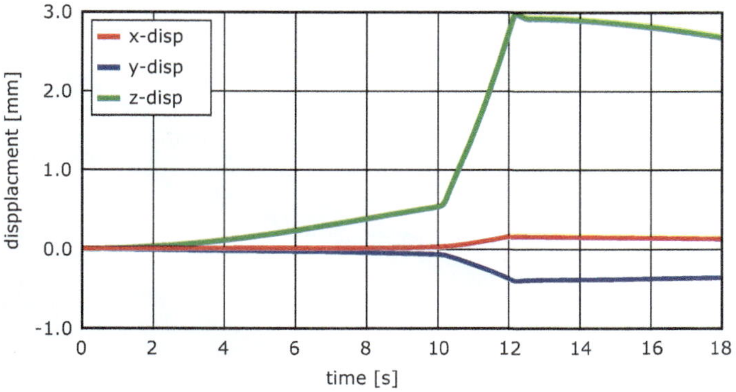

Fig. 10 Reduced order model manoeuvrer flight scenario

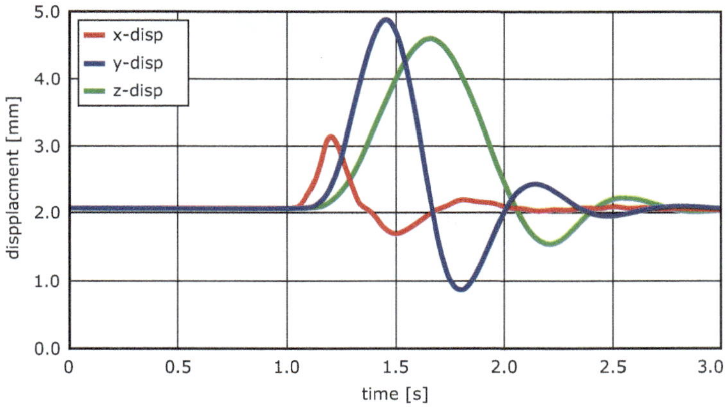

Fig. 11 Reduced order model manoeuvrer flight scenario

Here, v_g represents the gust upwind field, V is the gust speed, $v_{g,0}$ is the gust amplitude, and l_g is the gust gradient. Figure 11 shows the structural displacement response of the wing tip for different gust gradients.

Using the example of the take-off and gust encounter scenarios, critical load cases can be determined with small computational costs. In this case, a simplified aerodynamic strip model is used for the simulation. Depending on the accuracy requirements, this can be extended via higher order methods (see for example [12–14]).

The focus of this project is the determination of critical load cases. Because the transformation matrix is known, the internal loads of the structure can be transferred back to the full FE model and a detailed stress analysis can be performed for the critical condition. The displacement dependent strains are reconverted to a

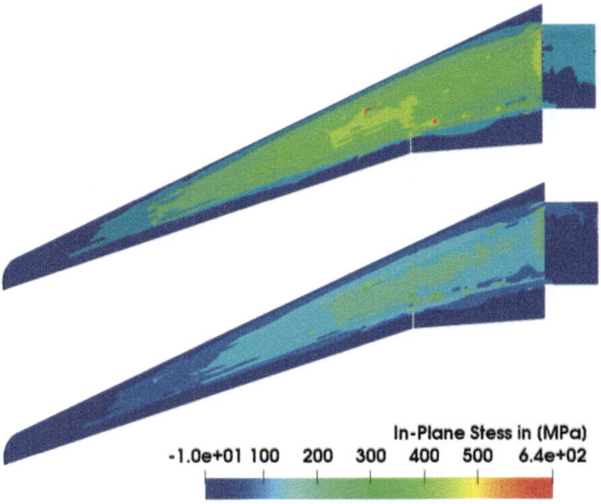

Fig. 12 In-Plane stress lower skin for two different gust load scenarios, upper small gust gradient, lower long gust gradient

detailed model, which reads in the matrix notation with the linear differential operator matrix [L]:

$$\epsilon = \left[L\right]\left[T\right]\mathbf{x}_{red} \tag{9}$$

With the constitutive law, the stresses are related to the strains, as shown exemplarily in Fig. 12, for a critical short and long gust length encounter load case.

4 High-Fidelity Elastic Aircraft Studies

The design of quasi-steady trimmed flight conditions with increased load factors n_z is discussed here. In general, such load cases contribute not only to the static and initial structural design, but also to the determination of trimmed flight conditions considering structural deformations. In VitAM-Flex, methods based on solving the RANS equations are applied to obtain the aerodynamic loading.

The lift polars for a rigid body and for an elastic wing are shown in Fig. 13, with the results for the two reference load cases involving manoeuvrer load factors of $n_z = 1.0$ and $n_z = 1.4$. The simulation is initialised with the jig-shape wing structure at an AoA of 0°. The solution steps of the trimming algorithm are indicated in red. As the number of iterations increases, the algorithm approaches an equilibrium state, which must lie on the elastic lift-polar curve.

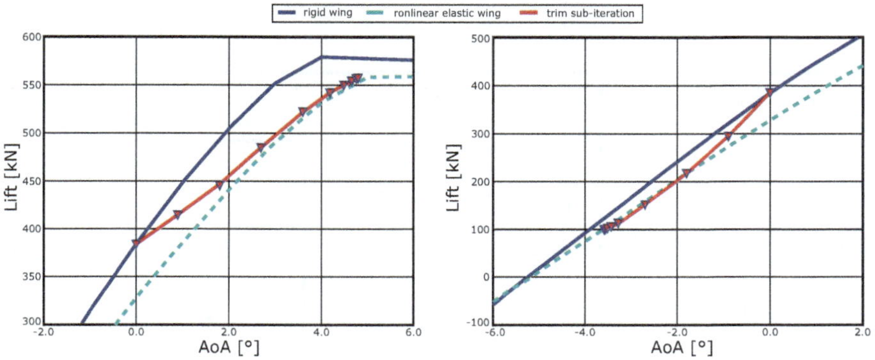

Fig. 13 Fluid interaction coupling scheme (left), structural eigenmode mapping scheme (right)

4.1 *Investigations on Flutter Analysis of Elastic Wing Structures*

The trim algorithm is further used to investigate the influence of wing deformation on the flutter speed predictions using the p-K method. The starting point of the investigation is the design of the flutter wing, as shown in Fig. 2. The case study is carried out with a cruise Mach number of $Ma = 0.8$ at flight level $FL = 330$. The necessary AoA is determined for a load factor of $n_z = 1$ under the quasi-steady-state flight condition. The resulting flight shape at $\alpha = 0.268°$ is shown in Fig. 14. The framework developed for the aeroelastic simulation is used to investigate the response around the nonlinear 1g trimmed flight shape of the wing.

The structural equation of motion described by means of the generalised coordinates q, is shown below:

$$\bar{M}_{gen}\ddot{q} + \bar{B}_{gen}\dot{q} + \bar{K}_{gen}q = F_{gen}. \tag{10}$$

Fig. 14 Wing deformation at 1g quasi-steady trimmed level flight condition at Ma = 0:8, FL = 330

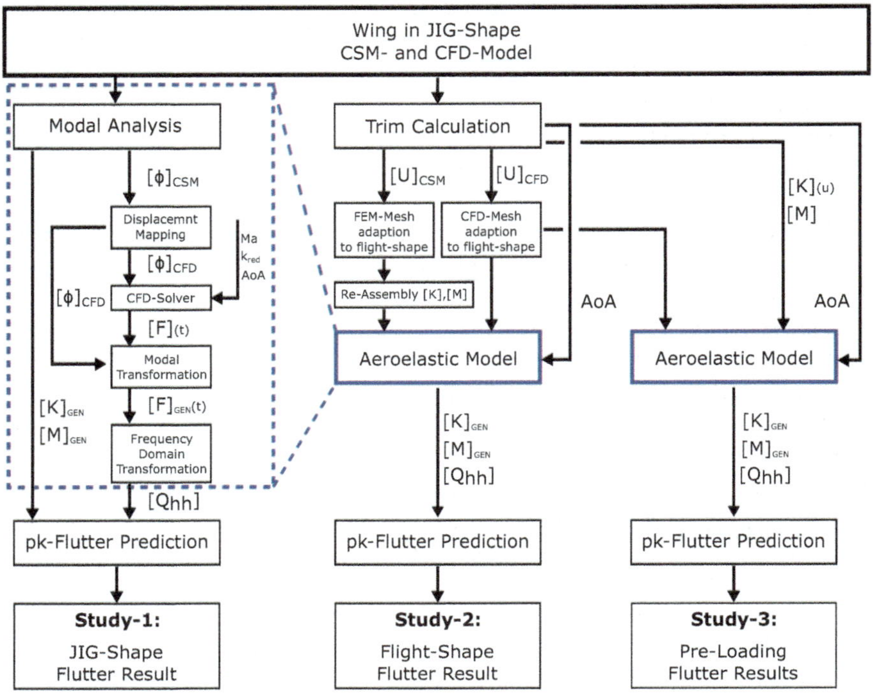

Fig. 15 Case studies to investigate the influence on the flutter prediction of preloaded wings

where $\bar{M}_{gen}$, $\bar{B}_{gen}$, and $\bar{K}_{gen}$ denote the generalised mass, damping, and stiffness matrices respectively [10]. Considering the effects of the stiffness and lift changes separately, flutter calculation according to the p-K method is implemented in the existing aeroelastic simulation framework. Three case studies are conducted to examine the influences separately:

1. Wing model in the jig-shape.
2. Wing model in the flight shape without structural preloading (only change in lift distribution is considered).
3. Wing model in the flight shape and structural preloading is considered.

The essential frameworks of the three simulation models and their sequences are shown in Fig. 15.

The generalised aerodynamic forces Q_{hh} for arbitrary flight conditions are determined based on CFD methods in the time domain and subsequently transformed into the frequency domain. The imposed structural eigenmodes Φ are scaled to an amplitude of $\frac{1}{500} \cdot c_m$, as suggested in [11].

Fig. 16 Pressure distribution of the wing's upper and lower surface, jig-shape (left) and flight-shape (right)

4.1.1 Aerodynamic Model Analysis

The pre-stress has no influence on the aerodynamic model; therefore, the changed wing shape is the only factor that affects the lift distribution and thus, the load vector. This is demonstrated by the pressure distribution between the jig- and flight-shapes, as shown in Fig. 16. Owing to the elastic wing, local wash-out (reduction in the AoA) occurs near the wing tip.

4.1.2 Structural Model Analysis

To evaluate the changes in modal stiffness, mass and frequency, the resulting static trimmed 1g level flight is incrementally loaded on the structure. The variation in the modal properties scaled with the magnitude of the unloaded structure for each mode and load increment λ is shown in Fig. 17. It can be observed that not all the modes are influenced by the preload in the same manner. Notably, mode-3 is dominated by bending and torsion mode-9 showed the largest changes in the modal properties.

To increase the stiffness resulting from the preload, the load increment is increased to $\lambda = 1.5$. The changes in $\lambda = 1.0$ and $\lambda = 1.5$ with respect to the stress-free state are summarised in Table 3. It can be observed that the influences on the modal mass and stiffness matrix increases. For the flight conditions at 1g, it can be summarised that no significant changes are recognisable; thus, only minor changes are reflected in the structural model used for the calculation of the critical flutter speed.

In addition to the changes in the natural frequencies, the natural eigenmodes are examined in more detail. The evaluation of the similarity between jig-, flight-shape and preloaded structures is performed by means of the MAC criterion, which is listed in Table 4.

The flight-shape wings with and without the preload show no significant changes ($MAC \approx 1.0$). However, larger deviations, especially for mode-8 and mode-9, can be identified between the jig- and flight-shape wings.

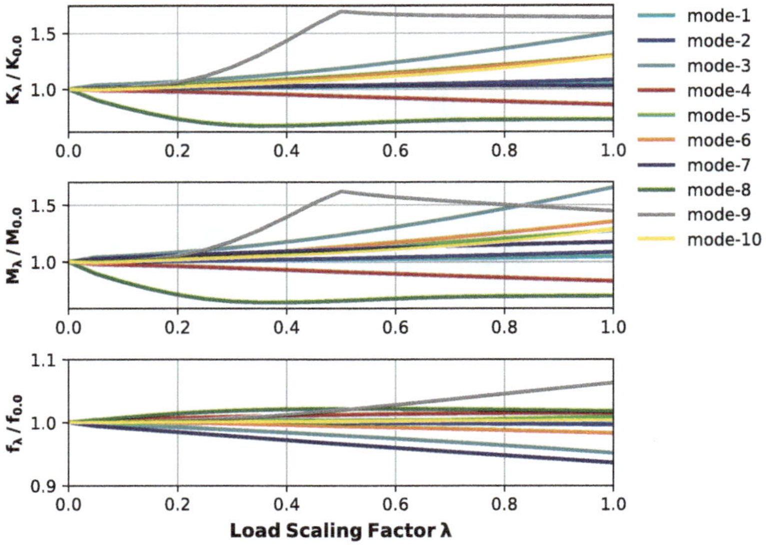

Fig. 17 Changes in modal properties due to external structural loading

Table 3 Changes in mass and stiffness matrix due to pre-loading effect. $\Delta\% = |\Delta K_{gen}|/K_{gen}$

λ = 1.0	**Mode-1**	**Mode-2**	**Mode-3**	**Mode-4**	**Mode-5**
ΔM_{gen}	0.06%	0.14%	0.90%	0.20%	0.08%
ΔK_{gen}	0.38%	0.03%	0.36%	0.08%	0.02%
	Mode-6	**Mode-7**	**Mode-8**	**Mode-9**	**Mode-10**
ΔM_{gen}	1.62%	1.59%	0.27%	2.96%	1.20%
ΔK_{gen}	1.29%	2.04%	0.24%	2.99%	1.28%
λ = 1.5	**Mode-1**	**Mode-2**	**Mode-3**	**Mode-4**	**Mode-5**
ΔM_{gen}	0.84%	12.84%	1.14%	12.21%	14.62%
ΔK_{gen}	9.47%	8.54%	5.05%	16.35%	13.04%
	Mode-6	**Mode-7**	**Mode-8**	**Mode-9**	**Mode-10**
ΔM_{gen}	13.34%	3.30%	0.42%	5.10%	37.60%
ΔK_{gen}	11.37%	0.15%	3.00%	3.88%	37.71%

Therefore, only the geometric shape change plays a role in the structural model for determining the flutter limit. Although, preload effects can be demonstrated, they only affect structural dynamic responses at higher load factors. This implies that, specifically for the 1g level flight, no additional computational effort is necessary to obtain a new set of generalised aerodynamic forces to assemble the load matrix Q_{hh}.

Table 4 MAC value results of eigenmode shapes comparison

Structural mode	1	2	3	4	5	6	7	8	9	10
jig- to flight-shape	0.987	0.975	0.972	0.976	0.968	0.951	0.980	0.484	0.100	0.962
Flight-shape to preloaded wing	1.0	0.999	0.999	0.998	0.999	0.999	0.998	0.999	0.998	0.999

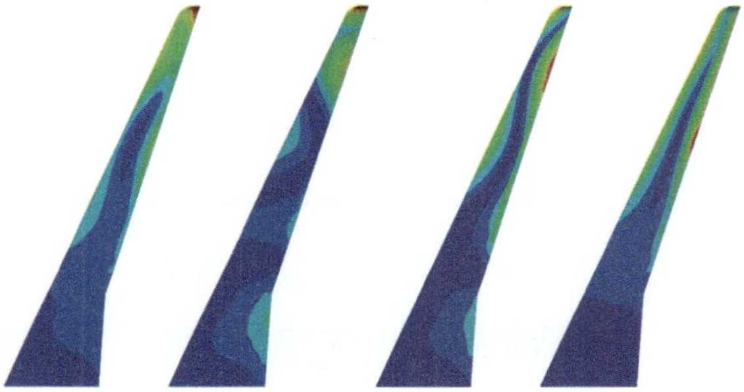

Fig. 18 Wing eigenmode shape comparision between jig- and flight-shape, mode-8 (left), mode-9 (right)

4.1.3 Aeroelastic Model Analysis

The results of the p-K flutter prediction method are shown as a root locus plot of the eigenvalues in Fig. 19. The influence between the flight-shape and the prestressed flight-shape structure is small. Significant deviations are observed between the jig- and flight-shape. For example, mode-4 in the jig-shape case is unstable and becomes non-critical with the wing's deflection to the flight-shape. The flutter limits for all the three cases are summarised in Table 5. In addition, the frequency and damping trends over the dynamic pressure q_∞ at $Ma = 0.8$ are shown in Fig. 20.

To maintain low computational costs, only a half-model with symmetrical boundary conditions is considered for the results presented. Thus, not all the physical effects required for a complete description of the aircraft are covered. However, it was found that important changes in the aeroelastic behaviour occur in the presented case and changing from the jig-shape to the flight-shape is associated with the strongest influence. Therefore, the flight shape should be used for aeroelastic designs.

4.1.4 Summary and Outlook

The work performed in VitAM-Flex helped to introduce design methods for the early structural pre-design process. Particular emphasis was placed on industrial applications; hence, a transport aircraft configuration with an increased aspect ratio was selected. The current trend in new aircraft configurations indicates configurations

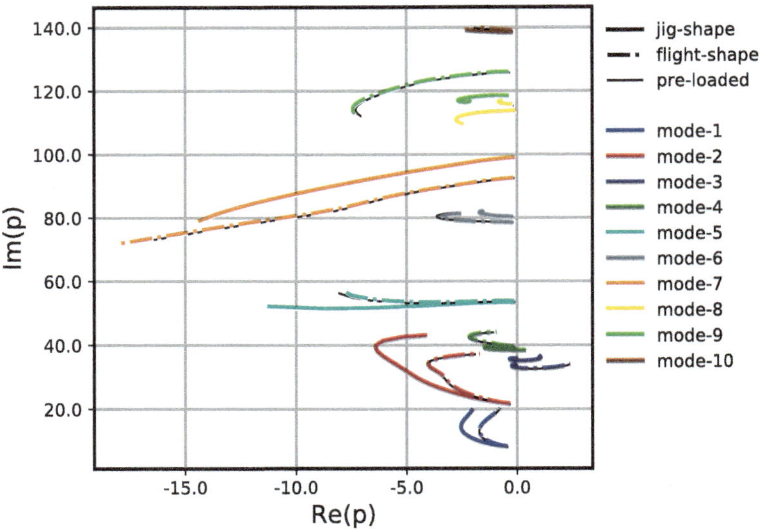

Fig. 19 Pressure distribution of the wing's upper and lower surface, jig-shape (left) and flight-shape (right)

Table 5 Flutter boundaries for the three case studies

	Jig-shape		Flight-shape	Pre-loaded
$q_{\infty,f}[kPa]$	Mode-3	Mode-4	Mode-3	Mode-3
	30.4	40.5	25.5	26.1
$k_{red,f}[-]$	0.150	0.141	0.153	0.152
$f_f[Hz]$	5.542	6.064	5.170	5.152

with increasing structural flexibility. Therefore, it is necessary to accurately represent important characteristics such as the coupling of different disciplines and underlying physical models in the virtual model.

In the first part, structural reduction methods were investigated to study elastic aircraft configurations effectively and quickly. The IRS method was identified as an adequate method for describing elastic wing structures. Furthermore, its application was demonstrated using a virtual take-off manoeuvre and gust-load encounter. Through this reduction, the mathematical structural model can be solved considerably faster, such that virtual flight manoeuvres for load evaluations or flight controller tests are possible. A purely linear kinematic consideration of this problem is critical. In particular, the methods investigated must be extended for large wing deformations.

In the second part, higher order methods were applied to the aircraft structural design process. Large nonlinear deformations were verified using strongly coupled CSM-CFD methods. In this context, the impact of large structural deformations on flutter predictions was investigated using the established industrial design methods.

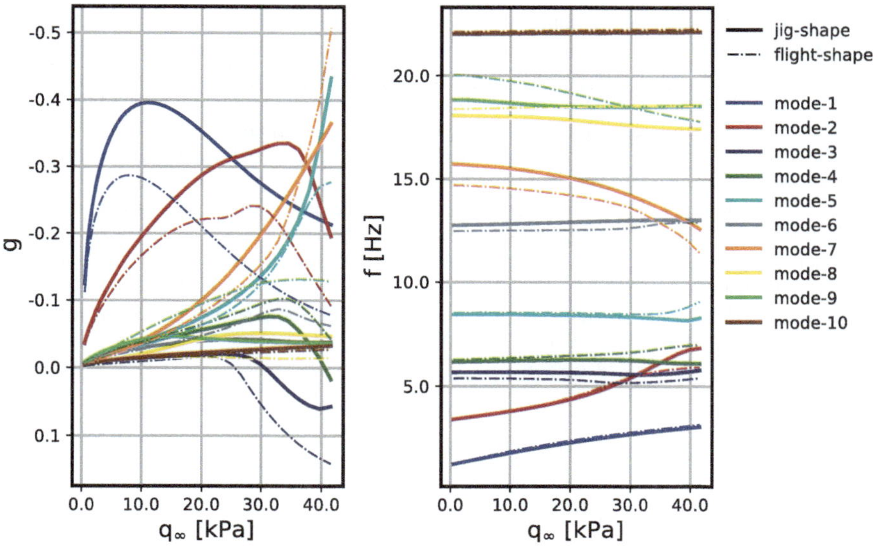

Fig. 20 Pressure distribution of the wing's upper and lower surface, jig-shape (left) and flight-shape (right)

A significant influence could be identified owing to changes in the aerodynamic lift and mass distribution. The results achieved in this project form the basis for further research, including UAV flight tests for validation.

References

1. Aerogust Consortium, Aeroelastic Gust Modelling, Final summary of the AEROGUST project. www.aerogust.eu
2. Görtz, S., Krumbein, A., Ritter, M., Hofmann, J.: DLR-Projekt VicToria—Virtual Aircraft Technology Integration Platform. Deutscher Luft- und Raumfahrtkongress, Friedrichshafen (2018)
3. Guyan, R.J.: Reduction of stiffness and mass matrices. AIAA J. **3**(2), 380 (1965)
4. Salvini, P., Vivio, F.: Dynamic reduction strategies to extend modal analysis approach at higher frequencies. Finite Elem. Anal. Des. **43**(11), 931–940 (2007)
5. Gordis, J.H.: An analysis of the Improved Reduced System (IRS) model reduction procedure. Int. J. Anal. Exp. Modal Anal. **9**(4), 269–285 (1994)
6. Friswell, M.I., Mottershead, J.E.: Finite Element Model Updating in Structural Dynamics, 1st edn. Springer, Dordrecht (1995)
7. EASA (2020) Certification Specification and Acceptable Means of Compliance for Large Aeroplanes CS-25. https://www.easa.europa.eu/sites/default/files/dfu/CS-25%20Amendment%2024.pdf
8. Breuer, M., De Nayer, G., Münsch, M., Gallinger, T., Wüchner, R.: Fluid-structure interaction using a partitioned semi-implicit predictor-corrector coupling scheme for the application of large-eddy simulation. J. Fluids Struct. **29**, 107–130 (2012)

9. Farhat, C., Lesoinne, M., Le Tallec, P.: Load and motion transfer algorithms for fluid/structure interaction problems with non-matching discrete interfaces: Momentum and energy conservation, optimal discretization and application to aeroelasticity. Comput. Methods Appl. Mech. Eng. **157**(1), 95–144 (1998)
10. Freymann, R.: Strukturdynamik, 1st edn. Springer, Berlin, Heidelberg (2011)
11. Fleischer, D.: Verfahren reduzierter Ordnung zur Ermittlung instationaerer Luftkraefte, Dissertation Technische Universitaet Muenchen (2014)
12. Albano, E., Rodden, W.: Doublet-Lattice method for calculating lift distributions on oscillating surfaces in subsonic flows. AIAA J. **7**(2), 279–285 (1969)
13. Sherwood, H., William, M.: Nonlinear Programming Extensions to Rational Function Approximation Methods for Unsteady Aerodynamic Forces, NASA Technical Paper 2776 (1988)
14. Hesse, H., Palacios, R.: Reduced-Order aeroelastic models for dynamics of maneuvering flexible aircraft. AIAA J. **52**(8), 1717–1732 (2014)

Printed by Printforce, the Netherlands